Biowaste and Biological Waste Treatment

Biowaste and Biological Waste Treatment

Gareth Evans

Published by James & James (Science Publishers) Ltd,
35–37 William Road, London NW1 3ER, UK.

© 2001 Gareth Evans

The moral right of the author has been asserted.

All rights reserved. No part of this book may be reproduced in any form or by any means electronic or mechanical, including photocopying, recording or by any information storage and retrieval system without permission in writing from the copyright holder and the publisher.

A catalogue record for this book is available from the British Library

ISBN 1–902916–08–5

Printed in the UK by The Cromwell Press

Cover photos courtesy of Tony Boydon and Talbott's Heating Ltd., A.R. Newton (Morawetz UK Sales) and Gareth M. Evans.

Contents

Foreword	ix
Preface	xi
Author's Acknowledgements	xiii

Chapter 1 – An Introduction to Biowaste   1
    Problems of disposal   5
        Leachate   5
        Methane   7

Chapter 2 – The Management of Wastes   10
    Analyzing Waste   10
    Management Options   14
        Biowaste Recycling   16

Chatper 3 – The Regulatory Framework   21
    Standard Industrial Requirements   21
    General Waste Laws   22
    Specific Biowaste Issues   23
    The UK, Europe and the Landfill Directive   24
    The American Situation   27
    Product Standards   29

Chapter 4 – Biological Waste Treatment   32
    Collection Regimes   36
        Mixed MSW   37
        Separate Collection Schemes   37
        Recycling Banks and Amenity Sites   37
    Marketing Issues   41
        Heavy Metals   42
        Health Risks   46
        Economic Factors   54

Chapter 5 – Composting   58
    The Four Phases of Composting   59
    Compost Biology   61
    Composting as a Method of Biowaste Treatment   63
        Home Composting   63
        Centralised Composting   66
        Screening Equipment   71

    Process Parameters   72
        Temperature   72
        Moisture Content   73
        Aeration   74
        Particle Size   74
        Nature of the Feedstock   74
        Processing Time   75
        Accelerants   76
        Maturation and Curing   76
    Measuring Compost Maturity   76
    Mulch Production   77
    Uses of Biowaste-Derived Compost   78
        Agricultural Applications   79
        Plant Disease Suppresssion   80
        Environmental Applications   82
    Health Issues   84
    Suitability for MSW   85

Chapter 6 – Anaerobic Digestion (AD)   89
    An Overview of the Main Processes of Anaerobic Digestion   90
    The Three Stages of Anaerobic Digestion and Methanisation   90
        Hydrolysis   90
        Acidogenesis   91
        Methanogenesis   91
    The Bacterial Ecology of Anaerobic Digestion   92
    The Chemistry and Production of Biogas   94
    The Role of Hydrogen   96
    Process Variables and Operational Conditions   97
        Physical   97
        Chemical   100
    Anaerobic Digesters   102
        Operational Digester Types   104
        AD Systems in Use   106
    Mass Balance and Product Utilisation   107
        Digestate Utilisation   108
        Biogas Utilisation   109
        Process Liquor: the Forgotten Product   111
    Suitability for MSW   114

Chapter 7 – Alternative Biotechnologies   118
    Annelidic Conversion (AC)   120
    Ethanol Production   124
    Eutrophic Fermentation (EF)   126
        Process Stages   128

Viability and Potential for Scale Up	131
Changing Circumstances	133

Chapter 8 – Thermal Recycling: Energy from Biowaste — 136

Energy in Biomass Systems	137
Paper: The Forgotten Biomass	139
Refuse-Derived Fuel	141
Woody Biowaste	142
Pyrolysis	142
Gasification	143
Short Rotation Coppicing	144
The Scope for Biowaste Product Use	145
Biomass and Renewable Energy	147

Chapter 9 – The Way Ahead — 150

Options for Landfill Diversion	152
Selective Ban	153
Limiting Actual Amounts	154
Tradable Permits	155
Developing Alternative Approaches	156
Taxation Issues	158
The Future of MRFs	161
Waste Minimisation	162
Sustainable Waste Management	164
Best Practicable Environmental Option	165
Integrated Resources Management	167

Chapter 10 – Policy and Planning — 169

The Biowaste Treatment Plan	169
Objectives and Constraints	171
Technology Selection	173
Intrinsic Factors	174
External Factors	176
Monitoring and Facility Performance Criteria	180
On Site Considerations	181
The Products	181
Normal Operational Considerations	181
Extraordinary Considerations	181
Evaluation and Analysis for Client and Contractor	183
Biowaste Treatment and Recycling Obligations	184

Contacts — 187

Addresses, telephone numbers, email addresses and websites of selected sources of information

Index — 192

Foreword

by Dr Caroline Jackson MEP

Chairman of the European Parliament's Committee on the Environment, Consumer Protection and Public Health.

This is a timely book on an unglamorous subject, but one which repays study. Action by the European Union, in the form of new legislation on waste, and especially through the 1999 directive on landfill, has brought the issue of waste disposal up the agenda dramatically. When the directive first appeared, I described the United Kingdom as 'one of Europe's last 'throw away' societies' because we send to landfill over 80% of our household waste.

All EU Member States are now committed to meeting the directive's timetable to reduce the amount of biodegradable waste going to landfill over the next 20 years. In the case of the United Kingdom, and other major landfill users such as Spain and Greece, the directive implies a major switch in policy from landfill to other forms of waste management. This is not only a huge challenge for local authorities and local planners, but also a test for government. It is small comfort that other developed and developing countries are grappling with the same problem.

In this book **Gareth Evans** provides a comprehensive account of the biowaste question, covering not just the technologies but the underlying practical issues of implementation as well. It is too easy for politicians – and others – to make out that our waste management problems can be solved simply by increasing recycling and opting for waste minimisation strategies. These will play a role, but they are only part of the answer. Biowaste and biological waste treatment must be taken into account and explored as we move away from over-reliance on landfill.

This book will be a useful resource for all those professionally involved – client and contractor – in waste management at this crucial stage in its development, and I commend it to the reader.

Caroline Jackson MEP

Preface

by the Rt Hon. Michael Meacher, MP
Minister for the Environment

The promotion of sustainable waste management is one of the Government's key environmental objectives. Our primary aim is to reduce the amount of waste which is generated and to manage the waste which is then produced by more sustainable means, such as re-use, recycling, composting and energy recovery.

The Government is determined to see an increase in the amount of waste composted and recycled, in order to meet the challenges of sustainable development and the stringent targets in the EC Directive on Landfill.

The Landfill Directive seeks to prevent or reduce possible negative effects on the environment by the landfilling of waste. It aims to ensure high waste disposal standards by setting rigorous controls on the operation of landfill and the emissions from these sites. Many of these controls are similar to those already in place in England for landfills.

To control emissions of methane (a powerful green house gas) from landfill, the Directive sets out challenging targets for reductions in the landfill of biodegradable municipal waste. These targets require us to landfill less and less of this waste over the next two decades to just 35% of the amount of such waste produced in 1995. Meeting these targets will require the diversion of substantial amounts of this waste to municipal waste alternatives, such as composting, recycling, energy from waste and other technologies. This represents a major step-change for the management of municipal waste in this country, and a challenge for local authorities, the waste industry and Government.

This challenge is a central theme of this Government's strategy for waste, which was published in May 2000. In this, the Government set two goals of recycling or composting 30% of household waste by 2010, and recovering 45% of municipal waste (by recycling, composting and incineration with energy or heat recovery). Thereafter we expect to recover value from two-thirds of household waste, half of this by recycling or composting,

Already, an increasing number of local authorities are setting up centralised composting schemes to compost waste from their own operations, from separate household collections and from civic amenity sites. Voluntary groups, too, are setting up community composting schemes to collect and compost organic material on a co-operative basis. And there is a growing trend towards encouraging individual householders to compost their own biodegradable waste.

The actions, targets and aims that I have outlined here, illustrate this Government's commitment to an integrated approach to waste management, and the move towards sustainable development as a whole.

Waste costs us all money. It is in the interests of every person, every business to cut the amount of waste we generate. Sometimes the simplest actions – for example redesigning a product to make better use of the materials from which it is manufactured, or to make it more easily recycled once its working life is over – can lead to substantial savings.

Government cannot achieve this radical change on its own. Everybody – every business, every householder, every community group and service provider – has a role to play in making sustainable waste management a reality across the UK. We must all become more waste-aware, and we must all start taking a more active, responsible attitude to the waste we all generate.

Author's Acknowledgements

I became involved in biological waste treatment largely by accident. As a child it had never really occurred to me to wonder what happened to the refuse we threw out every week. Although we had a compost heap of sorts, and I had a vague notion that our dustbin contents went to what was then called 'the dump', it never dawned on me that the two things were in any way related. Likewise, throughout my studies of biology, I do not recall ever being aware that a whole profession existed to make practical applications of biodegradation. For me, decomposition was something which happened on the forest floor, or at the bottom of the ponds frequented by the gulping toads and fragile newts that I loved so much. Having been plucked, not quite literally, from those amphibian infested waters, I now know differently.

The purpose of this book is to examine the present, and likely future, state of biological waste treatment. Something of a Cinderella science, this has at last begun to assume greater importance as an increasingly necessary part of modern waste management, driven, in no small part, by the burgeoning of stringent environmental legislation around the world. The discussion falls naturally into three parts. The first section (chapters 1–3) covers the nature of biowaste, waste treatment in general and the regulatory framework which governs it. The second (chapters 4–8) looks at the technologies and approaches available, while the final part (chapters 9 & 10) examines the various policy questions and local, social and economic factors which affect the implementation of biowaste initiatives.

Waste management is very context-dependent and a myriad of local influences can alter the picture from country to country, so while the benefits from any particular approach may need to be judged independently, the broader aspects of the core technology remain relevant. Thus, throughout, I have deliberately tried to avoid lengthy descriptions of individual methods of well-known, and generically available, technologies, concentrating instead on the wider issues which apply universally. The more unusual processes have merited greater detail, but, hopefully, only as much as is needed to advance the discussion, rather than to confuse the issue. Having spent a good number of recent years developing and promoting various biowaste processes for commercial use, principally in the UK, this experience has inevitably shaped my views. However, I have tried to overcome any tendency to bias by the inclusion of some of the huge amount of work, both laboratory and operational, done elsewhere, particularly other European countries and North America. I hope I have been successful in providing a fair picture. Equally, I hope proponents of landfill and incineration will not feel that their methods have been too harshly treated. Both have a role to play in waste management, but this is a book about biowaste and I do firmly believe that there are few occasions when either of these technologies are truly the most appropriate route for kitchen and garden refuse.

Many people have helped me along the way, in one way or another, and I am thankful to them all. A number of photographs appear by permission of their owners. I gratefully acknowledge the kindness of my friends Rob Heap and Bob Talbott, together with Blackwall Ltd, Morawetz (UK Sales), O Kay Engineering Services Ltd, TEG Environmental PLC, Tony Boydon Photography, Traymaster Ltd, and Vagron (Holland) in this regard. I must also thank Nora Goldstein and David S. Riggle for their assistance with the section on American regulations. In particular, I have drawn heavily on David's lecture *The Evolution of Composting and Compost Standards in the United States (An Overview of How the Composting Industry Has Developed and How and Why Standards Have Been Adopted)* which he presented in Hungary. I am grateful to him for letting me have the transcript and his permission to use it in this way. No work of this nature could be undertaken without reference to many sources and acknowledgement has been made of these, where they appear. However, thanks must also go to Gicom BV, Plus Grow Environmental Ltd, Portec Rail Products (UK) Ltd, Scarborough Environmental Services, and Vibraflo (BJD100 Ltd) who provided information which, though not specifically referred to in the text, was, never-the-less, valuable in helping to set the wider context.

I am also obliged to James and James, in particular to Edward Milford for giving me the opportunity to write this book to begin with, and to John Pacione and Alan Peterson, at the outset, and Gina Mance, latterly, for their guidance along the way.

I will finish by saying a public thank you, firstly, to my wife, for all of her help and support, not just during the writing of this book, when she regularly kept me supplied with coffee and read through the drafts, but throughout my biowaste career. Secondly, to my parents, for giving me the education which enabled me to enter this particular area of science in the first place and for encouraging me to enjoy writing.

Gareth M. Evans,
County Durham, 1st February, 2000

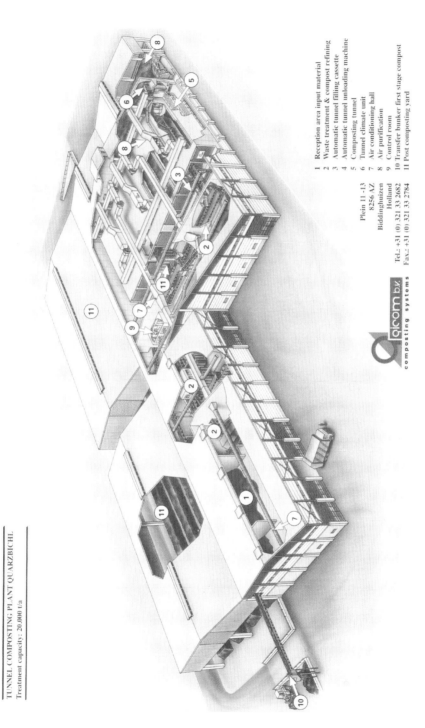

Plate I. Tunnel composting plant schematic; courtesy of Gicom B.V.

Plate 2. The Biomass Anaerobic Digestion pilot plant (showing dewatering gantry); photograph by the author

Plate 3. The Biomass Anaerobic Digestion pilot plant (showing heating system); photograph by the author

Plate 4 (above). Maturing digestate held in the hand; photograph by the author

Plate 5 (left). Windrow turner in action (front view); courtesy of Traymaster Ltd

Plate 6. Clean MRF; part of a system to process 100,000 tonnes of MSW per year; courtesy of O. Kay Engineering Services Ltd

Plate 7. Windrow and windrow turner; courtesy of A.R. Newton (Morawetz UK sales)

Plate 8. Biomass-fuelled 2,000 Kw/hour automatic boiler; courtesy of Tony Boydon and Talbott's Heating Ltd

Plate 9. Composting cage system; courtesy of TEG Environmental Ltd

Plate 10. Home composting bin; courtesy of Blackwall Ltd

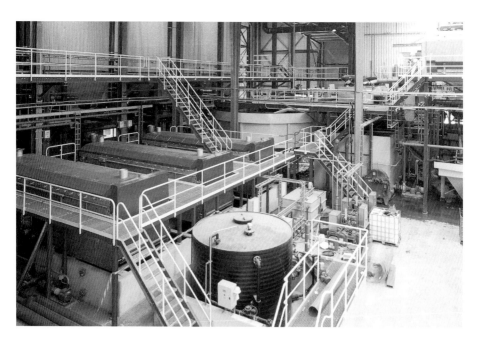

Plate 11. Interior of the new Groningen anaerobic digestion plant showing the washing plant on the right and the fermentation facility on the left; courtesy of Vagron

Plate 12. The four anaerobic digestion reactors of the Groningen plant; courtesy of Vagron

CHAPTER 1
An Introduction to Biowaste

As growing public pressure and burgeoning environmental legislation have led to the wider adoption of a hierarchical view of waste management options, the clear aim has emerged of developing more sustainable methods of dealing with refuse in general. The introduction of ever more stringent regulations on the waste industry by the relevant governmental authorities, coupled with moves to reduce reliance on landfill as a disposal route, makes it increasingly likely, particularly within Europe, that some form of biological treatment will soon become a standard requirement for the vast majority of putrescible waste.

In addition, since almost all such waste can be converted into a reusable product of one form or another, the biological treatment of biowaste has implications for any overall re-use/recover/recycle strategy, an issue which will be discussed in more depth later.

Historically, there has been some confusion over the terminology applied to wastes deriving from living or once-living sources. Hence, they have been variously described as *putrescible, green, food, yard, biosolid, garden* or, simply, *organic* wastes, largely depending on their origin and the regionally, or nationally, accepted usage of the person doing the classification. While this approach makes an apparently natural distinction between materials, it can equally serve to mislead. Does, for example, *food* waste include meat scraps along with the vegetable peelings, and if not, is it then truly synonymous with *green* waste? How much of *garden* waste is readily *putrescible* weeds and how much is woody prunings; how does British *garden* waste compare with its American *yard* counterpart?

To further muddy the water, the actual sorting protocols employed to obtain data on waste arisings can themselves introduce errors of representation, by an over adherence to a particular system of pre-defined classifications. Thus, for example, it is only within the last ten years that official UK analysis has split its traditional 'putrescibles' into 'kitchen' and 'garden' wastes, with Warren Spring Laboratory additionally introducing an extra 'fines' class into their recording process, which, defined as less than 10 mm in size, may often have a high organic content. Clearly, attempting to make any sensible comparisons between different management regimes or different countries' published data for waste production is not helped by categories which allocate broadly the same material into specifically different groups.

The term *biowaste*, then, used to encompass all of the above kinds of materials and more, into a single class, while expressly rejecting the root-of-origin criteria of classification, encourages the development of a process-based approach to their systematisation. Hence, wastes tend to be considered in terms of their ease of decomposition or treatment, falling into more natural classes and sub-classes on this basis, rather than being shoe-horned into essentially artificial pigeon-holes. The derived biowaste taxa which arise as a result are, moreover, more useful and directly relevant to the consideration of appropriate technologies for their ultimate processing or utilisation.

Biowaste arises from a variety of human, agricultural, horticultural and industrial sources and can be considered to form three general groups; waste of directly animal origin (faeces/manures), plant materials (grass clippings and vegetable peelings) and processed materials (food industry and abattoir wastes). Despite the earlier criticism of root-of-origin categories, this classification is entirely consistent with the process-based approach advocated previously, since the members of these groups all share certain fundamental characteristics and hence naturally decompose in essentially similar ways.

From a purely chemical standpoint, biowaste may be considered to be characterised by a high carbon content, though in practical terms this, of course, effectively means substances derived from fresh living matter. This is an important distinction, since synthetic organic wastes from fossil fuels and their residues, or chemically treated plant or animal materials, often prove resistant to biological breakdown. The decomposition of untreated biological material has formed the basis for the literal recycling of nutrients and chemicals in nature for millennia. However, these natural cycles have been somewhat disrupted by growing human activity, particularly during recent history, as the following consideration of ancient and modern mass flows per capita[1] clearly shows.

Table 1.1 Mass Flows Per Capita (kg/person/year)

Substance	**Ancient**	**1988**
Carbon	50	2,200
Nitrogen	5	40
Phosphorus	1	0.4

Returning now to the three broad groups of biowaste, as defined earlier, it should be obvious that the amount and composition of animal-origin wastes produced will show considerable variance. This depends on a number of factors, including climate, body size, age, diet, daily intake and in the case of livestock, management regime, species or breed, while cultural and socio-economic influences affect the situation in humans.

Thus, a 65kg person may produce between 0.1 and 0.5kg of faeces per day, and 15–35g of Biochemical Oxygen Demand (BOD), a 400kg cow, between 18 and 40kg of manure (500–900g BOD) and a 2kg laying hen excretes 0.04–0.06kg with 4g BOD. While such figures are interesting in themselves, it is important to remember that, at least from the point of view of biowaste treatment, it is not the gross physical amount, nor entirely its biological or chemical oxygen demands that form one of the major constraints, but rather the total available nutrients. Moreover, the full implications of this are often easy to overlook, as the following tables illustrate.

While, once again, the figures themselves are of interest, it is sometimes more relevant to consider these animal-origin wastes on a weight for weight basis. If the same figures are adjusted to reflect the nutrient production of each of the three species per kilogramme of body weight, the view becomes somewhat different, as in table 1.3.

Table 1.2 Approximate Annual Nutrient Production (kg/year)

Nutrient	Human	Cow	Hen
Nitrogen	8.5	55	0.7
Phosphorus	2.8	15	0.6
Potassium	2.1	20	0.3

Table 1.3 Annual Nutrient Production (kg/kg body weight/year)

Nutrient	Human	Cow	Hen
Nitrogen	0.13	0.14	0.35
Phosphorus	0.04	0.04	0.3
Potassium	0.03	0.05	0.15

Though the mammals' contributions are closely similar, weight for weight, it is obvious that poultry manure is of a quite different character, which goes some way to explaining both the specific problems and the particular applications associated with this kind of biowaste.

Since one of the desired main objects of modern biowaste treatment is to generate a usable and, ideally, saleable, final product at the end of processing, an awareness of available nutrient potential is inescapable. However, for all such wastes, the emphasis must rest squarely on the word 'potential', since in a practical application, the nutrient levels actually available for plant uptake may often be less than a straightforward analysis of the raw materials would indicate. Though controlled biodegradation can be very effective in transforming large, complex organic molecules into smaller, simpler constituents, the leaching of nutrients either during processing, or *in situ*, once applied to the soil, coupled with the sometimes inevitable inability of particular crop species to utilise the nutrients in the form in which they are presented with them, can make the maximum yield difficult to realise.

Biowaste derived from plant material itself is principally composed of cellulose, though with differing amounts of other plant structural compounds, including hemicelluloses and lignin. Cellulose itself is a structural carbohydrate made up of a complex chain of hexose units. While hemicelluloses superficially resemble cellulose, they are more readily hydrolysed; lignin, however, found in woody tissue, xylem vessels, sclerenchyma and strengthening fibres, is the most complex of all the plant structural carbohydrates and, though being originally derived from the conversion of cellulose, this material is much less readily decomposed. To put this into context, the biowaste fraction of Municipal Solid Waste (MSW) in the United Kingdom typically consists of around 40–50% cellulose, 5–10% hemicelluloses and 10–15% lignin

Plant material is of routinely high water content, often 80–90%, and while this may require to be taken into consideration during processing, in general terms, plant-derived biowaste, like the preceding class, is eminently suited to biological treatment regimes. However, in the urban context, the large amount of grass clippings seasonally available can pose its own set of problems for biological waste management initiatives, both at the household-composting scale and for larger, central facilities.

Waste stabilisation is fundamentally brought about by the decomposition of the expressly biodegradable material, together with the associated reduction of the carbon to nitrogen (C:N) ratio. Clearly, the two things are linked, since the degradation of the biological material proceeds by converting significant quantities of carbon to biogases (principally carbon dioxide and/or methane, dependent on the method of treatment being employed) which in turn lowers the C:N ratio, as required. Since a C:N ratio less than 18:1 is widely taken to be necessary to achieve product stabilisation and a C:N ratio higher than 25:1 will actually begin to inhibit the desired mineralisation of nitrogen, it should be apparent that for optimal processing, careful control of the input material will be essential. This aspect is of particular importance, since it is this plant-derived class of biowaste which is increasingly tending to be seen as the easiest means by which Local Authorities and the like may meet their present and future statutory obligations, particularly in respect of the targeted diversion of such wastes from the landfill disposal route.

Clearly, much of the waste from horticultural sources, together with an appreciable amount of agricultural crop residue wastes, also fall into this group. With growing interest being shown from many quarters in on-farm composting schemes, there is conspicuous potential for further developing the synergy of such a combined treatment and utilisation option.

The final category, that of processed materials, is something of a catch-all, essentially more diverse than either of the proceeding 'natural' classes. There are, at least from a chemist's viewpoint, certain wastes which, though they fit the wider definition of 'organic' used by that discipline, like plastics, leather, rubber and manufactured woods (MDF and Conti-Board) for example, never-the-less do not truly belong here. Inherent in the concept of biowaste is the need to be biodegradable. While that has been defined as a 'material in which all organic ingredients are degraded completely in biologically active environments to simple molecules . . . carbon dioxide and/or methane and water, leaving behind only biomass and naturally occurring assimilation products'[2], there is also an implicit, if undefined, time limit. Biowaste is readily biodegraded, a point picked up by the consignment of much of these non-biowaste organics to the combustible fraction by certain authorities.[3, 4]

Thus, though the chemical constituents of discharges arising from facilities as different as food processing plants, abattoirs, pulp mills and breweries are obviously quite dissimilar, they all represent biowaste. While their individual BOD, COD, pH and suspended solids content may all vary enormously and the treatment regimes required to render them adequately innocuous for final disposal differ, biological processing, in one form or another, remains relevant to them.

Biowaste represents a huge fraction of society's refuse; the European Union alone produces some 2,500 million tonnes annually[5] and this figure is growing by the year. There is a tendency to focus on the biowaste element contributed by MSW, since this is both the kind of waste with which everyone is most familiar and the one for which the responsibility for disposal normally rests with a local authority, rather than on the householder. This, however, tends to distort the true picture. Of the 2,500 million tonnes of EU biowaste mentioned, the bulk is agricultural (1,000 million tonnes) with garden and forestry wastes accounting for 550 million tonnes, sewage sludge 500 million, food processing 250 million and MSW a mere 200 million.

There is, then, a very large amount of biowaste around. Based on figures from a Swedish study, it has been estimated that there is between 850 and 1,000 kg (total solids) of biowaste suitable for biological treatment produced per capita, per year[6]. However, it remains the case that the potential for the development of these biological treatments has not yet been systematically investigated, nor have their possible applications been greatly explored. The coming years seem likely to change this, as the ever increasing demands of environmental legislation drive fundamental changes in the way waste is viewed and treated. Consequently, as moves around the world to reduce the amount of untreated biowaste entering landfills gain momentum, processes which deal both effectively and acceptably with this material must surely begin to assume a more central role in the waste management pantheon. Their integration into the wider spectrum approach that will be required to meet the likely stringent requirements of forthcoming laws offers exciting possibilities and real challenges to the industry as a whole. Ultimately, biological treatment technologies may come to assume the kind of reverential status currently reserved for traditional materials recycling.

Such things lie in the future, but the beginnings are already apparent. The European Landfill Directive and various other laws in Germany, Holland, the US and elsewhere have started the move. The writing is on the wall for our end-of-pipe, out of sight, out of mind, disposal paradigm. When a third of dustbin refuse is biowaste, and thus consigned to landfill, or incinerated for its dubious calorific contribution, how much longer will we really be able to justify recovering the mere half percent represented by aluminium beverage cans and still claim the environmental equivalent of the moral high ground?

Problems of Disposal

The disposal of biowaste in landfill also raises the question of pollution, since as water percolates through, it leaches out both organic and inorganic compounds which may leave the site via the subsoil or site drains and culverts. This is, clearly, undesirable both on amenity grounds and because of its potential effect on ground water. The persistence of pathogens and the edaphic translocation of many chemicals of biological activity have recently become growing concerns for environmentalists, waste managers and public health officers alike, with the mounting (though, admittedly, currently largely circumstantial) evidence of health problems associated with certain landfill operations.

Leachate

Even for those sites where there is no suggestion of a directly adverse effect on human health, high levels of biowaste-derived leachate remain something to be avoided, since organic-rich liquors offer a ready food supply for heterotrophic micro-organisms. Where the total organic loading thus available is relatively low, a balance point can be reached fairly readily between the bacteria breaking the material down and the autotrophes, typically algae, utilising these breakdown products and a dynamic equilibrium is established between them. This works because, at the low end, the oxygen required for breakdown is offset by the oxygen produced by the algae present. However, in contrast, when the organic loading is high, the oxygen demand of the dissimilatory bacteria exceeds the carrying capacity of the water and the algae's ability to replenish it. Hence a downward spiral, rather than an equilibrium, develops, ultimately leading to locally anaerobic conditions.

It is worth noting in passing, however, that, under controlled conditions, exactly these interactions have formed the functional basis of certain of the technologies used to bring about the treatment of biowaste, though they rely heavily on effective process monitoring and careful control of the reactor environment to achieve their goal.

To begin to put the pollution issue in some kind of context, landfill leachate analysis ranges are commonly quoted based on the average values obtained from established sites. This can tend to distort the picture since, particularly for newer tips (where the processes are more acetogenically fermentative than 'old' post-methanogenic or even semi-aerobic) the levels of some parameters appear under-represented.

This point is made in the following table, by comparing 'typical' acetogenic leachate ('early stage') with methanogenic ('late stage')[7].

Table 1.4 A comparison of early and late stage leachates

Determinand	Acetogenic	Methanogenic
COD	22,000	3,000
pH	6.1	8
Sulphate	500	80

pH expressed in pH units, concentrations in mg/l

Acetogenic leachate is, thus, characterised by a pH of <7 and a high COD, though much of the latter is biodegradable. Since the exact complement of bacteria responsible will vary with environmental conditions, they may be anaerobic, aerobic, or facultative anaerobes. Methanogenic bacteria, by contrast are strict anaerobes and will only appear in the first place and continue to thrive in the absence of oxygen. In the normal run of things, as in a mature landfill, their dominance of the microbial population comes as a result of acetogenic bacteria having used up the available oxygen (though, while not relevant for this discussion, their acetate substrate production also plays a part in this).

Table 1.5 Real-life landfill ranges

Determinand	Sample 1	Sample 2
BOD	1,413	35,100
COD	3,926	45,600
pH	7.56	5.83

pH expressed in pH units, concentrations in mg/l
Sample 1 Deponie Bassum (Germany) PCI Membrane Systems Ltd
Sample 2 Stewartby (UK) Centre for Environmental Research and Consultancy (CERC) study

Furthermore on this particular topic, methanogenic leachates are of lower COD, but a very high proportion of this may not be degradable. Table 1.5 shows the kind of real-life ranges by means of data from two different landfill sites.

Moreover, the Centre for Environmental Research and Consultancy (CERC) study [8], of which the Stewartby figures formed part of a much larger investigation, produced a general range for landfill leachate:

BOD	80,000 – 5
COD	100,000 – 50
Nitrate	1,000 – 0.1
VFAs	50,000 – 150
Sulphate	1,000 – 1

pH expressed in pH units, concentrations in mg/l

Methane

A second aspect of biowaste-derived pollution from landfills, and perhaps a more worrying one in global terms, is the production of methane (CH_4), which is, as a greenhouse gas, more than thirty times more damaging than similar amounts of carbon dioxide (CO_2). The mechanisms of methanogenesis will be discussed in greater detail in a later section, but suffice it to say at this point, that in a landfill, particularly deep within it, where the ingress of oxygen is prohibited, large organic molecules are converted into methane and carbon dioxide by the action of bacteria. Under ideal anaerobic conditions, this ultimately proceeds to yield fully reduced methane and fully oxidised carbon dioxide.

The reality of this breakdown at the microscopic level is chemically very complex, involving hundreds of potential intermediary reactions and compounds. In addition, many of these each have further need of specific synergistic chemicals, catalysts or enzymes. In very general terms, however, it is possible to simplify the overall biochemical reaction to:

$$\text{Organic material} \longrightarrow CH_4 + CO_2 + H_2 + NH_3 + H_2S$$

It must be borne in mind, though, that some organic materials, like lignin for example, as discussed earlier, do not breakdown readily in this way, nor, obviously, do non-organic inclusions within the waste.

Concern regarding the methane generation from landfill sites was one of the initial driving forces behind the European Union's original desire to enact binding statutory controls to reduce the amount of putrescible biowaste material destined for disposal by this route. The initial proposal had been for the reduction of methane generation by the banning of wastes with a Total Organic Content (TOC) of more than 10% by weight from landfill within five years of the appropriate legislation being passed. Accordingly, a number of Member States, most notably France and Germany, unilaterally adopted a self-limiting target in line with this 10% TOC figure. The final version of the regulation, which reflects a great deal of the national politics which surrounded its stormy eight-year passage and which will be examined in more depth later, stops short of actually making this law. However, the pollution issue is alternatively addressed by requiring all sites, except those taking inert waste, to employ leachate collection and comply with stated universal minimum specifications for containment liners, while landfill gas must be collected and either used for energy generation, or flared where, for whatever reason, this is not possible.

It is perhaps worth noting, as we begin the next millennium, that biowaste is our most fundamental form of waste product. Though for our nomadic remote forebears, and even for our more direct ancestors of the Neolithic, this would have raised few waste management problems, other than making sure that discarded food remains did not lead would-be predators right up to the threshold of their cave. However, with the advent of the agrarian lifestyle, as settlements developed and grew, and the population density increased, this picture changed and has continued to change with each successive human advance. Indeed, it has been suggested that many of the worst methane problems associated with biowaste under traditional landfill regimes can be attributed to alterations of lifestyle and, ironically, at least in the UK, to some of our early attempts to address pollution concerns[9].

Hence, the provisions of the Clean Air Act, 1956 established Smoke Control Areas, the nett effect of which was to reduce the cinder and ash content of domestic refuse markedly. The presence of significant quantities of these inert spacer materials in landfill has been linked with the easy diffusion of produced gases to the atmosphere, thereby diluting the problem and avoiding the pooling of methane and its mass underground migration familiar today. This is not to suggest, of course, that the indiscriminate release of a gas with so powerful a greenhouse effect is to be recommended, but it does place the matter in historical context. Moreover, the 'new' practice of capping landfills, which came into vogue at about the same time, although successful in its design purpose of reducing the production of leachate, by preventing the passage of water through the deposited material, tended to exacerbate the lateral migration of landfill gas. Coupled with political changes, which reduced the number of local authorities while enlarging the responsibilities of the remaining ones, the result was the production of very large landfill sites, the internal conditions of which effectively guaranteed that biowaste would become more problematic than ever before.

A cursory glance over the kinds of waste consigned to dustbins today compared with, say, the 1950s would reveal a considerable change in our habits, marked by the profusion of plastics and aluminium beverage cans, not to mention the increased volume, of our modern waste. Advances in technology and materials are, inevitably, reflected in what we throw away and the waste of tomorrow may contain items as different again. Future laws, perhaps banning multi-material packaging, for example, may change our refuse; lifestyles may alter, but the single unifying thread between this hypothetical future and our remote past is the inescapable nature of our biology. Unless the 1960s' predicted nightmare of meals reduced to a single nutrient-rich pill comes to pass, we will always be tied to eating and that, inevitably, means peelings and food scraps, together with all the agricultural and horticultural biowaste produced by those industries which feed us.

For a number of reasons, some of which will be explored more fully in the next chapter, biowaste in general, and the biowaste fraction of MSW, in particular, seem set to receive greater attention over the coming decades. While, as mentioned earlier, the contribution of MSW-derived biowaste to the whole is relatively modest, it remains an area of considerable interest to local authorities and waste management companies charged with dealing with it. The manner in which this material currently arises and is collected brings certain complications to the situation, which will need to be resolved as part of any generalised strategy for its treatment, but these are not insurmountable obstacles, given the growing political will and public awareness. However, before moving on to a consideration of the biological treatment options available, the following two chapters will first examine the underlying rationale and regulatory framework of biowaste management.

References

1. Frostell, Dr Bjorn M., *The Role of Biological Waste Treatment in Integrated Waste Management*, Proceedings of Biowaste '92, ISWA/DAKOFA publication, 1992, p. 4, quoting A National Environmental Declaration Medborganas Miljomanifest, AB Timbro Stockholm, MOU 1990:2, pp. 160–185.
2. DHV Environment and Infrastructure BV (Amersfoort)., in cooperation with Plancenter Ltd, (Helsinki) and University for Soil Management, (Vienna) *Composting in the European Union*, a final report to the European Commission DGXI, Environment, Nuclear Safety and Civil Protection, 1997, p. 22.
3. Frostell, Dr Bjorn M., *The Role of Biological Waste Treatment in Integrated Waste Management*, p. 7.
4. IEA, *Biogas from Municipal Solid Waste*, An Energy Recovery from MSW Task Anaerobic Digestion Activity publication, 1996.
5. Bert Lemmes, *The 'Tao' of Organics*, Wastes Management, The Monthly Journal of the Institute of Wastes Management, September 1998, p. 18.
6. Frostell, Dr. Bjorn M., *The Role of Biological Waste Treatment in Integrated Waste Management*, p. 8 and personal comment.
7. Morris, Steve, *Landfill Leachate Generation & Physical Characteristics*, UKPS Ltd/PCI Membrane Systems Ltd.
8. Cope, Dr C. B., *An Introduction to the Chemistry of Landfill Gas and Leachate*, Centre for Environmental Research and Consultancy (CERC).
9. Cooper, George, *Anaerobic Digestion*, Introduction to the report of the Institute of Waste Management Anaerobic Digestion Working Group, May 1998, p. 9.

CHAPTER 2
The Management of Wastes

Analyzing Waste

Municipal Solid Waste (MSW) is a descriptive phrase frequently to be heard being bandied about in discussions and, while universally understood in broad terms, like a number of expressions within the waste industry, the precise definition tends to change dependent on the geographic origin of the speaker. Hence, in the UK, MSW tends to be used as exclusively synonymous with 'domestic' or 'household' refuse; in some parts of Europe, a broader definition of 'municipal waste' is generally applied, which encompasses that material which, though originating from commercial or industrial sources, is essentially similar to household waste in character and composition. This has caused some difficulties for the European Union's attempts to standardise waste legislation, since implementation depends on interpretation, which itself depends on national definition, at least in the absence of EU ruling to the contrary.

Moreover, it is widely appreciated that to begin to formulate anything resembling a rational response to the problem of waste management, at the level of either local strategy or national policy, a sound understanding of the composition of the refuse in question is essential. Here again, MSW will be found to vary from country to country, from region to region and even between different areas of the same town. Back in April 1992, when the provisions of the UK's Environmental Protection Act, 1990 were first coming into force, the idea that household refuse varied according to socio-demographic criteria was viewed as near-heretical in many established quarters.

At this time, changes in certain aspects of the manner in which waste was regulated in the UK began to open the way for new specialist companies and novel technologies for the treatment of refuse. Accordingly, for those seeking to offer a customised and integrated solution to an individual authority's waste management needs, on a local basis, the then common assumption that household waste throughout the country was the same, became untenable. A more comprehensive awareness of the particular vagaries of a given waste stream became inevitable and consequently, some local authorities, seeking to explore the full potential of the upcoming options, commissioned their own studies, while, on a national basis, the analysis work at Warren Springs Laboratory and the University of East Anglia assumed a new importance.

However, one of the problems with any programme of waste analysis is that, if it is to be directly relevant to a given treatment or disposal route, it must reflect the requirements of that route, *a priori*, in its own formulation. Hence, the categories, level of detail and accuracy required to make effective consideration of any particular option, are largely dependent on the type of waste treatment being examined[1]. Thus, to return to the UK example, when Halton Borough Council announced its initial intention, in September 1992, to be the first to follow a particular route based very heavily on a new biowaste treatment approach, in order to facilitate the design of the proposed plant, a small pilot study was made of its waste. Interestingly, as well as showing some variance between waste in Halton and Birmingham, according to a near-contemporary investigation there, the results also suggested that it was socio-demographic considerations which defined the waste profile, rather than the method of collection. The study areas were chosen to include examples of high-income and low-income areas and collections by both traditional black-bag and the newer wheeled bins. In all cases, income bracket proved to be the deciding factor in compositional analysis[2].

Furthermore, even when essentially similar techniques are being considered, different organisations and countries have applied differing criteria, which inevitably leads to the production of data of limited value, and the necessity for much work to be repeated, in order to meet individual requirements. In this way, taking the earlier example from Europe, less is known about the industrial and commercial wastes which approximate to household refuse than truly 'domestic' MSW in terms of their composition and it is, therefore, not immediately obvious from what data are available, how much of this would be suitable for biological treatment, for instance. There is a clear need for the establishment of more universally agreed standards of categorisation, sampling and analysis, particularly if the transfer of technologies between countries is to be facilitated. Until then, it remains necessary to make reference to the standards adopted and the methods used, in order to attempt meaningful comparison.

Notwithstanding, there is, fundamentally, one basic approach in general use for the estimation of municipal solid waste fractions. This method is largely site-specific and involves obtaining samples of the given waste stream, sorting out the individual components, and then weighing them. This approach gives a reasonable 'feel' for a given waste stream and can also provide useful insights into seasonal and climatic variation, the effects of population density and demographics. Moreover, for certain types of waste, and biowaste falls into this category, only by this kind of directly measurable approach can any truly meaningful estimate of arisings be made. However, while this method is very helpful in characterising a local waste stream, particularly if the regime involves the taking of a relatively large number of samples over a lengthy time period, and ideally over several seasons, it is not without its disadvantages.

For one thing, the statistically limited number of samples, even in a relatively large study, has the potential to encourage the drawing of misleading inferences, with certain categories possibly either over, or under, represented accordingly in the final results. Atypical circumstances encountered during the investigation, for example unseasonably wet or dry weather, a one-off delivery of an unusual waste type, or even errors in the execution of the sampling and analysis itself, could all give rise to major errors when multiplied up to represent a community's entire annual waste stream.

It should also be obvious that when a limited number of sample events, irrespective of whether they originate from single or multiple study locales, are factored up to provide a national picture, the potential for error magnification becomes much greater and considerably more significant. Though this kind of approach is very good on a local basis, the costs involved in properly extending it to the production of country-wide estimates would be prohibitively expensive.

An alternative method for waste characterisation, which utilizes an altogether different approach, based on material flow, was pioneered in the late 1960s and early 1970s by the United States' Public Health Service and subsequently developed by its successor, the Office of Solid Waste at the Environmental Protection Agency Their current database represents over 20 years of evolution and refinement, which, incidentally, overcomes one of the other criticisms of the straight-forward sampling approach in that, unless applied consistently and for fairly lengthy periods, it does not generate much in the way of useful data on trends in the waste stream over time.

The material flow method makes use of production data for the materials in the waste stream, relying on a series of modelling assumptions and adjustments which take into account such factors as the generation rates of the different commodities and anticipated product lifetimes. Further fine-tuning is employed to allow for imports and exports and diversion from the waste stream by way of recycling[3]. However, even under this approach, biowaste assessment is still dependent upon findings from direct sampling studies, though the inevitable anomalies discussed earlier for this kind of survey are to some extent mitigated by the use of the widest possible variety of individual investigations. Though this approach, by concentrating largely on manufacturing rates, takes no account of any residual contents which remain inside containers made from other materials, like sauce in a bottle or paint in a can, it is, generally, the most suitable means for estimating the waste arisings on a national basis.

Nevertheless, to reiterate the earlier point, the most appropriate method of assessment, the categories, level of detail and accuracy required, will always need to reflect the intended end use of the results, if it is to be of any direct assistance in helping to understand the specifics of the given situation.

Despite an almost universal acceptance of the need for accurate data and that, in the words of the UK Government's 1995 White Paper, 'good information is essential to the formulation of sound waste management policies', obtaining it seems to remain a widespread problem. The Environment, Transport and Regional Affairs Committee's Report[4] stated:

'The continuing lack of information in Government about waste is extraordinary: it would appear to be common sense that one first identifies the nature and scale of the problem before attempting to sort it out. The production of accurate statistics on waste arising, the composition of waste at the point of arising and on the demographic structure of households (which affects that composition) must be a Government priority.'

There is little doubt that the statistical database for municipal solid waste is somewhat poor, a fact which has been commented upon by the European Commission in respect of all the Member States. This failing is, moreover, particularly pronounced in the sphere of discerning trends over time, since historically, there has been neither the driving interest, nor the centralised recording structures necessary to make such a large undertaking feasible.

Within the UK context, the essential first step towards providing a framework for consistent data collection on waste arisings and management was the replacement of the existing patchwork of Waste Regulation Authorities with the Environment Agency. This body was established as the single waste regulator, assuming also the pollution control function previously performed by the National Rivers Authority and thereby uniting both aspects of waste licensing under a single umbrella. This also helped to overcome some of the historic reasons for the poor statistical base, since the responsibility for regulating and managing wastes is otherwise diverse, usually resting with local authorities, which are subject to greater financial constraints on what they can achieve in the way of data collection.

As discussed earlier, the lack of an agreed, standardised classification system to govern waste characterisation has traditionally complicated attempts to define, or even describe, 'typical' arisings in a wholly consistent manner. In addition, waste is generally not routinely weighed by either producers or carriers nor, frequently, is it, at the point of disposal. Hence, it has always proved difficult to derive a reliable weight to volume ratio for modelling or statistical purposes. A Government review of waste handling procedures indicated that 24% of local authorities in England and Wales typically weigh none of their waste loads, while nearly 80% of the remainder (60% of the total) weigh less than half. According to Scottish Office figures, Scotland fares a little better, with around 67% of municipal solid waste being weighed prior to disposal. However, it is, consequently, inevitable that errors will arise in estimating the quantities of municipal solid waste produced nationally, despite the information that Waste Collection Authorities provide, in good faith, to the relevant surveys.

Table 2.1 Typical Categorisation of Municipal Solid Waste in the UK

Category	% by weight [a]	% by weight [b]	Inclusions
Paper and card	33.2	30.7	Newspapers, magazines, other paper, liquid containers, card packaging & other card
Biowaste	27	33.7	Garden waste, kitchen waste, other putrescibles & fines (<10mm)
Glass	9.3	7.9	Brown, green, clear & other glass
Miscellaneous combustibles	8.1	5.2	Disposable nappies & other combustibles
Dense plastic	5.9	3.4	Clear & coloured beverage bottles, other bottles, food packaging & other dense plastic
Ferrous metal	5.7	7.5	Beverage cans, food cans, other cans, batteries & other ferrous
Plastic film	5.3	4.6	Refuse sacks & other plastic film
Textiles	2.1	3.3	Textiles
Miscellaneous non-combustibles	1.8	2.5	Miscellaneous non-combustibles
Non-ferrous metal	1.6	1.2	Non-ferrous beverage cans & other non-ferrous metal

Sources:
[a] Warren Springs Laboratory data, presented to 1994 Harwell Waste Management Symposium
[b] An Introduction to Household Waste Management, ETSU for the DTI, March 1998

The estimates which do exist for the UK's household waste have been drawn from the National Household Waste Analysis Programme, which characterises households on the basis of socio-economic factors, which are reflected in purchasing patterns and lifestyle and, thus, in the character, composition and amount of waste produced. Sample collections from waste which has been previously identified as originating from required socio-economic groups are taken for sorting and analysis, with other influences, such as the relevant waste management practices of the local authority, being considered during sample selection. Coupled with the newly established targets for improving the accessibility and certainty of information, and increased monitoring of recovery, recycling and diversion from landfill, it seems likely that the availability of reliable data will improve in the future.

Bearing in mind the limitations of the data, table 2.1 gives an indication of the general composition of Municipal Solid Waste for the United Kingdom, based on two official sources.

Though the broad agreement is clear, discrepancies of even 2 or 3 per cent, in the context of the whole country's arisings of some 27 million tonnes, roughly one tonne per household, are significant amounts of waste. Moreover, for the purposes of planning the waste management needs for the future, a 2 per cent error represents over half a million tonnes of actual, physical refuse unaccounted for – the equivalent of more than 500,000 households' annual output, or that of ten typical local authorities. The need for accurate data is obvious, and the Government and the Select Committee are, clearly, absolutely right to afford it such priority.

Management Options

Traditionally, MSW has been dealt with in one of two ways: incineration or landfill. Though changes in the general perception of the whole issue of waste have brought other options into consideration, it remains the case that, on a world-wide basis, the vast majority of household refuse is still ultimately consigned to one of these disposal routes. Various countries have favoured one or the other option over time and while a lengthy discussion of either of these is beyond the scope of this work, a brief comparison of a few national waste management arrangements, as in table 2.2, may be of some use in establishing the wider context.

It is, perhaps, worth noting in passing that, the UK's contemporary heavy reliance on landfill notwithstanding, the first fully functional waste incinerator was built in 1874 in Nottingham, England[6], and operated for more than 27 years. Eleven years after the Nottingham facility was built, the world's first waste-fired electricity generating plant was opened at Shoreditch, London, in response to a growing awareness of the potential benefits of energy recovery. The technology has been refined and improved over the years, chiefly in other countries and a number of the original developers, like Vølund (Denmark) and Von Roll and Martin (Germany), remain major players in the energy-from-waste field today.

However, modern mass burn incineration, even with energy recovery, remains a disposal route, just like landfill. Increasingly, the 'higher' elements of the waste hierarchy, *reduce*, *re-use* and *recycle*, are being widely accepted as the way of the future, as part of a larger move towards a more sustainable approach to the whole of resources

Table 2.2 A Comparison of Selected Countries' Waste Management Arrangements

Country	Incineration (%)	Landfill (%)	Recycling (%)
Austria	11	65	24
Belgium	54	43	3
Canada	4	67	29
Denmark	55	20	25
Finland	4	66	30
France	33	59	8
Germany	36	46	18
Holland	42	30	28
Japan	74	21	5
Norway	18	68	14
Sweden	47	34	19
Switzerland	47	11	42
United Kingdom	9	85	6
USA	15	61	24

Source: IEA Bioenergy[5]

management. From the point of view of biowaste, it is hard to see how the waste minimisation philosophy could be applied, since we will always need to eat and changes in lifestyle, life expectancy and the generalised increase in our leisure time have made gardening one of the top pastimes of our age. Biowaste, it seems, is here to stay, if not actually increase. Much more potential exists in the areas of re-use and recycling.

Modern waste management best practice, as a whole, is moving away from the old fashioned, *burn it or bury it* view and this is particularly relevant for biowaste, since it is ideally suited to neither of these traditional avenues. For an incineration regime, its contribution to combined heat and power generation has been likened to trying to burn water; in landfill, the twin spectres of leachate and landfill gas loom large. The need for an integrated approach to waste treatment, with a very real emphasis on the 'treatment' aspect, rather than mere disposal, has gained clear recognition and stimulated a number of governments to set far reaching targets for the short to mid term. Additionally, legislative and fiscal moves have been instigated to further drive the transition away from the historical exploitative approach, towards a new paradigm of sustainability.

The principle of charging each producer of waste on the basis of the environmental impact of that material, and the method chosen for its treatment or disposal, would intuitively seem to be a fair and efficient way to promote waste minimisation and the selection of less harmful management routes. However, this is less clear-cut for household waste in general, and biowaste in particular, than would be the case, say, for industrial or commercial waste, where the audit trail is more easily established. Nevertheless, the introduction of the UK landfill tax, for example, fits well with the hierarchical view that sees reducing waste production in the first place as preferable to re-use, and any form of disposal only appropriate as a last resort, when no other beneficial use can be made.

Biowaste Recycling

The low likelihood of biowaste reduction has been touched on before. Hence, the recycling of biowaste represents the most productive means by which to mitigate the contribution of this particular material to the wider waste management question. In part, this has been directly addressed by the establishment of various composting and other biological treatment targets, by individual state or national legislatures, which is an area to be examined more closely in the following section on the regulatory framework. However, the rationale of biowaste recycling goes further than these directly prescribed quotas, set by relevant laws for the material itself. As the demand for increased recycling grows, and if the overall goals are to be met, then it is essential that the biowaste fraction be considered. Indeed, to contemplate high levels of overall recycling without a major contribution from the 30% or so of MSW represented by this material, must be fundamentally flawed. The recycling figures given previously in table 2.2, for example, expressly include composting.

However, as the following table illustrates, there is a considerable difference between how much biowaste exists in any given waste stream, how much it is realistically possible to take out for treatment, and how much actually is currently being recovered.

Table 2.3 Biowaste in MSW and Biowaste Recovery in the European Union

Country	Recoverable biowaste x10^3 t/y	Actual recovery rate (%)	Biowaste in MSW (%)
Austria	2,200	50	27
Belgium	1,670	19	47
Denmark	900	55	37
Finland	700	10	33
France	14,500	3	
Germany	9,000	45	30
Greece	1,650	0	50
Holland	2,000	90	
Ireland	350	0	36
Italy	9,000	2	25
Luxembourg	50	14	44
Portugal	1,200	0	35
Spain	6,600	0	44
Sweden	1,500	16	38
UK	9,240	3.4	30
EU Total	**60,560**	**15**	**36**

Source: European Commission DGXI[7]

Clearly, while the idea of recycling biowaste into useful and environmentally sound products has received almost universal support, there a number of factors which act to limit the implementation of such treatment options. Firstly, in the market-based approach, business development is inevitably driven by the real, or perceived, needs of the consumers. In theory, this should doubly benefit the cause of biowaste recycling, since opportunities exist in both the disposal of waste (a service industry) and the creation of a usable final product (a manufacturing industry). In practice, though, the situation often falls from this highly-desirable one of 'win–win', to a 'lose–lose', as the perceived market is seen as too small to warrant large investment by the operators, while, to open up the full potential for the large-scale users of composts, mulches and soil improvers, the material is simply not available in large enough quantities. It is a classic case of the chicken and the egg. Neither can be achieved without the other, and so neither is achieved.

Allied to this is the cost, both of the treatment itself and to potential customers, on either the waste disposal, or the final product, side. The price of both of the latter will be a critical factor in the marketing plan and if the key to the whole business venture is that it is a low-cost option for biowaste disposal, then the existing price of the alternative route will always be the commercially defining point, irrespective of any 'feel-good' factor that may be an attractive adjunct. The situation with regard to the finished product may offer slightly more leeway for the influence of perceived environmental advantages, however. While some potential users will still have to look to their bottom line, others, particularly local authorities, may be prepared to pay a modest premium for the 'green' kudos gained, particularly if this forms part of either an overall strategy, or a wider reduced-cost package.

The danger of this is that, at least initially, the unit cost of biowaste treatment facilities must almost invariably be higher than the existing disposal alternatives and, consequently, when stacked against them, the biowaste plant will always appear disadvantaged. This is, of course, an unfair comparison, for five main reasons. Firstly, the application of new technologies typically leads to an apparently disproportionately high first phase price, which, when replicated over time, tends to drop rapidly in real terms. Secondly, the direct comparison of new and exclusively biological treatments with current disposal methods is not a fair one, since, particularly in the case of landfill, the lifetime of existing facilities is often limited. A biowaste plant is limited only by the continued desire to keep it running; it has no void space to use up. Thirdly, while current options meet current needs, they may well not address the future legislative and political trends. Fourthly, such facilities may have other benefits to a local community by way of job creation or amenity value that are not immediately obvious from a straight-forward, price-per-tonne comparison. Finally, and it is a factor which must be close to the heart of all local authorities, recycling, and the rational husbandry of resources in general, increasingly represents the explicit will of the people.

Public perception, however, remains one of the whole issue's most mixed of blessings. While the advantages and innate 'rightness' of biowaste treatment have widespread appreciation and support, the practical matters of achieving it are still dogged, to some extent, by misconceptions. We may laugh now at the awful confusion '*recycled* toilet rolls' caused a few years back, but things have not entirely moved on as far as we might like to believe. Otherwise perfectly rational human beings express total dismay at their belief that waste-derived compost will contain significant quantities of disposable nappies, for example, completely missing the point of the required sanitisation processes prior to marketing. It has also been quite sincerely argued that, with a UK national average of 2.4 children, every household must throw away an inordinate quantity of diapers over the year, which will flood any centralised biowaste treatment facility. Sadly, adherents of this view have, of course, failed to remember that most of us give up wearing nappies long before we first attend school, which affects their calculations, rather considerably. In the face of the current level of public awareness and perception of biowaste-derived products, it seems likely that a targeted programme of information and education will be central to any successful expansion of biological waste treatment.

Legislation, or more accurately, the interpretation and application of it, can also serve to limit the cause of biological waste treatment. Within the UK, this has been described as 'an issue that affects the production of compost more than the potential outlets.'[8] The problem is exacerbated since there is little clear guidance regarding which aspects of biowaste treatment fall within the ambit of planning and which should form part of the waste management licensing procedures, nor are there any plans to issue such guidance on the specific issues affecting biological treatment. Clearly, both regimes have their part to play in ensuring that environmental and public health safeguards are upheld, and the local planners and the Environment Agency are specifically required, under the terms contained in Waste Management Paper 4, to act in a co-ordinated manner in examining proposals for waste treatment facilities. However, there will always be the potential for friction between local concerns and national policy; this is not a phenomenon unique to the UK.

Limiting Factors

Poor availability of facilities to act as examples of best practice may also be a limiting factor in the advancement of biowaste treatment. This is, of course, particularly relevant in the case of novel technologies, but can play almost as important a role for proven, but little used, treatments and even for well established, though locally under utilised, methods. It asks a lot of a local authority or other body charged with the rational, cost-effective management of waste, effectively to take a gamble on what they may perceive as an essentially unsubstantiated option. The problem is, perhaps, epitomised by the difficulties encountered by any new approach attempting to obtain licence assent. Without a competent track record, a licence is not easily forthcoming, but without this licence, the facility cannot be run, to establish the requisite successful history. It becomes, all too readily, a circular argument, going nowhere and the issue is only partly mitigated by systems of 'exemptions'.

The opportunities for decisions makers to visit an operational plant, or at the very least, receive testimonials from satisfied customers, inevitably carry much weight. It is, then, hard to see how any fundamentally new approach to biowaste treatment can succeed, initially, as a commercial operation under the current system. The only likely route to acceptance would appear to be the construction of the new plant expressly as a demonstration facility, to be run on this basis until such time as a satisfactory work record can be demonstrated and the effectiveness of the process exhibited beyond doubt. This, clearly, makes significant demands on any developers of, or investors in, novel biological treatments and stands as one of the major obstacles to the wider proliferation of biowaste recycling. In the UK alone, since 1992, when the changes brought about by the Environmental Protection Act, 1990 came into effect, a number of companies who had burst onto the scene intending to capitalise on the opportunities thus newly presented, have either abandoned biowaste research in favour of limited traditional recycling, or disappeared completely. Each one that has failed to live up to its promise has made the path that much harder for those who would follow. There is a clear need to back up the excellent research that has been done with hard cash, to ease scientific breakthrough into the market place. Sadly, there has been a generalised unwillingness to do so, which contrasts markedly with other spheres of biotechnology, where a more long term view on investment has been the norm.

Even for well-established biological treatment methods, like composting, the need to remain competitive with other existing waste management options acts as a natural ceiling on the potential income to be derived from gate fees. Consequently, since income from sales of the product has been widely seen as critical in maintaining the financial viability of projects, investment in research and development of new techniques and products has been severely limited.[9] If the trend of general price increases for the alternative routes continues, the operators of biowaste facilities will have more leeway to charge a higher initial gate fee. By thus reducing their dependence on sales revenue, greater investment may be made possible in the areas of product and technology, thereby enabling maximum advantage to be taken of secure, long-term outlets which are, in the final analysis, the key to sustainable diversion from landfill or incineration. What is clear, is that when landfill is relatively abundant and cheaply available, the take-up rate for recycling options in general, and biowaste treatments in particular, remains poor, as can be instantly seen from a comparison of the figures for Holland and the UK in tables 2.2 and 2.3.

There is, then, also a case to be made for some form of governmental intervention, either by way of subsidy, tax advantage or mandatory obligation, to support existing facilities and stimulate the development and expansion of biowaste technologies. It has been suggested from certain parties within waste management itself that a low landfill tax, and small intended annual increments, do little to advance the cause of sustainable approaches to the problem, while, equally, sending a very weak message to the industry in general as to the real priorities for future improvements and investment. This is a factor directly affecting waste management companies themselves. It may also be necessary for some kind of carefully targeted mechanism, perhaps akin to a widening of the scope of the concept of industry sector fiscal neutrality, which is currently favoured in relation to energy usage and CO_2 generation, to producers of waste and local authorities, to penalise or reward, according to their efforts.

However, there are limits on what local authorities, or anyone else, can achieve in isolation and it is unrealistic to imagine that all waste management needs can be addressed wholly within their essentially artificial boundaries. While within the European Union, for example, by law, the proximity principle, requiring waste to be dealt with as close as possible to its point of origin, and the best practicable environmental option (BPEO) must be given full consideration, individual circumstances need also to be taken into account. Accordingly, there has been a shift towards viewing future waste planning more on the basis of regional sustainability and this trend seems likely to gain momentum . Certainly, by removing some of the elements of unnecessary duplication, it would seem to offer the best means for local authorities in a given area to balance the costs involved against environmental impact, thereby minimising many of the current and future negative aspects of waste management practice. Additionally, this approach should also enable regional initiatives to be established which could help circumvent many of the factors mentioned earlier, which currently act as barriers to the expansion of biowaste treatment.

References

1. Barton, Dr John R. (University of Leeds) personal comment.
2. Evans, Dr G. M. and Evans, Dr C. L., *Halton Borough Council Waste: An Initial Study*, Biomass Recycling Ltd internal document, 1993.
3. United States Environmental Protection Agency's website, *http://www.epa/gov/epaoswer/osw/*
4. The Environment, Transport and Regional Affairs Committee's Report, 1998, Recommendation 2, Paragraph 20.
5. IEA Bioenergy: Task 23, N. M. Patel, AEA Technology Plc, Harwell, Didcot, Oxfordshire, OX11 0RA
6. Hering, R. and Greelly, S. A., *Collection and Disposal of Municipal Refuse*, McGraw Hill, New York, 1921
7. DHV Environment and Infrastructure BV (Amersfoort)., in cooperation with Plancenter Ltd, (Helsinki) and University for Soil Management, (Vienna) *Composting in the European Union*, a final report to the European Commission DGXI, Environment, Nuclear Safety and Civil Protection, 1997.
8. *Report of the Composting Development Group on the Development and Expansion of Markets for Compost*, Department of the Environment, Transport and the Regions, July 1998, section 6.1, p. 17.
9. Ibid., section 7.1, p. 18.

CHAPTER 3
The Regulatory Framework

The treatment of biowaste is typically regulated within a framework of legislation which acts at three main levels:

- Standard industrial requirements
- General waste laws
- Specific biowaste issues

Although it is the latter category which is of principal concern to this discussion, before moving on to examine it in greater depth, it is worth giving some brief consideration to the former two. For the sake of brevity, this will largely be confined to the UK situation, while acknowledging that similar ground is covered by other legislation, elsewhere.

Standard Industrial Requirements

In common with most commercial or industrial facilities, biowaste treatment plants are subject to the usual requirements regarding siting and planning, health and safety, fire regulations, considerations of vehicle movements, hours of activity and the like. In the UK, the Town and Country Planning Act, 1992, is the main legislation, though there is some overlap with the Environmental Protection Act, 1990 (EPA) and associated regulations, which impose additional conditions on waste management sites, consideration of which are typically run as a parallel adjunct to planning.

The Health and Safety at Work Act, 1974 applies to all work settings, requiring employers to protect, as far as reasonably practicable, the health and safety of their workforce and any others who may be affected by their operations. It also requires the production of a written health and safety policy, which should, of course, address areas of known potential risk, particularly in respect of waste handling procedures, even though biowaste itself falls outside of the scope of the Control of Substances Hazardous to Health Regulations, 1994, (COSHH). Under the Management of Health and Safety Regulations, 1992, the employers' duties contained in the older Health and Safety at Work Act are clarified and extended, particularly regarding the potential risks to which workers and others may be exposed. Accordingly, all work activities must have specific risk assessments made and companies with more than four employees must record these assessments in writing. The object is to identify potential hazards and, where possible, remove them; where this is not possible, action should be implemented to minimise the threat they pose. In the case of biowaste, the areas which have been shown to require consideration under these Regulations include the character of the waste itself, possible contaminants within it, plant machinery and equipment, airborne dust, biological aerosols, and on-site traffic movement.

General Waste Laws

Within the laws relating to waste in general, the provisions for site approval and proper operational management procedure inevitably grade into the requirements of wider industry, as described above. Thus, the consideration of potential locations for biowaste facilities should take into account the relevant Waste Local Plan, while centralised treatment plants will require to be licensed by the Environmental Agency, for which planning permission is an essential pre-requisite. There are only limited exemptions to this procedure, which, as was mentioned in the preceding chapter, can lead to problems for the commercial development of novel technologies, under certain circumstances. In some of the states of the US, on-farm and other relatively small biowaste facilities, together with sites treating certain closely defined types of low-risk materials, are subject to a far less onerous licensing process than is the case for the larger plants which are destined to receive thousands of tonnes of mixed MSW yearly. For example, the Californian Compost Operations Regulatory Requirements, which have been in force since 30th June, 1995, categorise all that state's composting facilities into one of five tiers, as summarised in the following table, with corresponding licence conditions based on their potential environment impact.

Table 3.1 California's Compost Operations Regulatory Requirements[1]

Excluded	Notification	Registration	Standardized	Full Permit
Homes, parks, schools, universities etc treating less than 500 cy/year	Composting operations which treat agricultural material, or less than 1,000 cy/year of green material	Green material facilities which compost between 1000 and 10,000 cy/year	Green material facilities which compost more than 10,000 cy/year	Mixed MSW composting facilities
Sites composting green or animal wastes from agricultural sources disposing of less than 2,500 cy/year	Researchers composting less than 2,500 cy/year	Animal waste plants processing up to 10,000 cy/year		

cy/year = cubic yard per year. 1 cubic yard = 0.764m^3

Section 74 of the UK Environmental Protection Act requires the applicant for a waste management licence to be a *fit and proper person*. To qualify, the applicant and any other relevant personnel must demonstrate that they have no convictions for offences which would have bearing on the application, possess adequate funds or have made suitable financial provision to meet the obligations imposed by the licence terms and are *technically competent* to operate the proposed site. Certificates of Technical Competence (COTC) are required for biowaste treatment plants and are obtained by satisfying the requirements of the Waste Management Industry and Training Advisory Board (WAMITAB). However, an operator who does not yet have such a certificate, but who has applied to WAMITAB for certification, will be deemed 'technically competent' for two years, allowing time for the procedures to be completed, provided that a satisfactory assessment is made by the Environmental Agency.

This *fit and proper person* condition is a very important aspect of the licensing regime, since once planning permission is granted and it has been established that no harm will result either to public health or the environment, under Section 36-3 of the EPA, a waste management licence must be issued to an applicant who is a fit and proper person, as defined by the Act.

A number of attempts have been made to establish targets for waste minimisation, diversion or recycling. In the UK, the 1995 White Paper *Making Waste Work: A Strategy for Sustainable Waste Management in England and Wales*, set out a number of general aspirational goals, including the recycling or composting of 25% of MSW by the millennium and 40% MSW recovery within five years after that. But these targets enforce no requirement for the rate of recycling on individual local authorities, being set nationwide and have no weight in law. Indeed, it is expressly stated in the White Paper itself that 'because of local circumstances some authorities will find it difficult to reach the target while others will go beyond it.'

Similar things have happened in the US, though with the difference that the recycling or diversion provision actually appeared in the relevant legislation of forty-three of the states. Though the targets have undoubtedly stimulated progress, it is much easier to pass law setting a goal than it is to achieve it, particularly when very few of the regulations have any teeth to encourage compliance and almost all are, again, not set for the individual municipalities, but at the state level. As with the UK, if any given authority fails to achieve the required level, there may well be some embarrassment, but for the most part, that will be all. In seven of the forty-three states, by contrast, it is mandatory for the relevant local authorities to meet the targets, with California, for example, having recourse to the imposition of fines of up to $10,000 per day.[2]

Much of the rest of what could be termed general waste legislation relates to the responsibilities and restrictions on the collection, transport and final resting place of the material. Clearly, this relates to biowaste, since these are common requirements which apply across the board. Provision for the manner and duration of holding waste prior to treatment is often made in the planning or licensing process, though aspects of it may appear elsewhere also, as in the US Part 243 Guidelines, which specifically seek to reduce the 'nuisance and to retard the harborage, feeding, and breeding of vectors.'[3]

Specific Biowaste Issues

The most significant regulatory controls on biowaste will always be those which relate specifically to it, since they establish the final word as to what may be done, while additionally possessing the potential to be a significant driving force for the uptake of biological waste treatment itself. Within the European Union, this role now falls primarily to the Landfill Directive; in the United States it is the Part 503 Rule and its derivatives.

The UK, Europe and the Landfill Directive

The passage of the European Union Landfill Directive was a particularly stormy one, and its final adoption by the Council of Ministers, on 26th April 1999, represented the culmination of nearly ten years deliberation over a means to reduce the amount of putrescible material entering landfills. Subsequently published in the EU Official Journal, it came into force on 16th July of the same year and all Member States must transpose it into their respective national laws by no later than its second anniversary, in 2001. Furthermore, Article 5 of the Directive requires the setting up of 'a national strategy for the implementation of the reduction of biodegradable waste going to landfills', which must be in place by the same date in 2003.

The initial proposal had been for the reduction of methane generation by the banning of wastes with a Total Organic Content (TOC) of more than 10% by weight from landfill within five years of the appropriate legislation being passed. Accordingly, a number of EU countries, most notably France and Germany, have adopted a self-limiting target in line with this 10% TOC figure.

However, during the Directive's lengthy passage through its various stages, both the concept of TOC and an approach that limited methane generation were dropped in favour of a phased reduction of the total quantity of biodegradable waste allowed. The final version is a three-phase approach for the reduction of organic loadings in municipal landfills. Accordingly, taking 1995 levels as a baseline, Member States are required to reduce the landfilling of 'biodegradable municipal solid waste' to:

- 75% by 2006
- 50% by 2009
- 35% by 2016

However, in order to accommodate the objections of countries with a heavy traditional reliance on landfilling (UK, Eire, Spain, Portugal and Greece), the facility of 'optional derogation' has been provided. This is something of an exercise in political expediency, allowing the already lengthily debated Directive to proceed without further delay from its principal opponents. Accordingly, Member States which landfilled 80% or more of their municipal waste in 1995, or the latest year before this, for which adequate data are available, are thus allowed to opt to defer the target dates for their compliance by up to four years. Thus, the revised schedule allows for an extension for each stage to 2010, 2013 and 2020 respectively.

As well as stretching the timescale for these objectors, (the 75% first target was originally set for 2002) the targets themselves have likewise been weakened. The first version of this approach used 1993 figures as the baseline and required an extra 10% final reduction, all to have been achieved by 2010, a full ten years earlier than a derogating country like the UK will now be permitted to achieve.

Nevertheless, this is very largely consistent with a number of the existing UK Government's waste strategy targets, which were set out in the 1995 White Paper *Making Waste Work*, namely:

- The recycling or composting of 25% domestic waste by 2000
- Home composting in 40% of homes with gardens by 2000
- Central composting of one million tonnes of waste by 2001
- 40% recovery of municipal solid waste (MSW) by 2005

Unlike these White Paper targets, however, the EU targets now set, have statutory force and under the terms of the European Union, all member state must incorporate them into their own national legislation within two years of their formal adoption.

Though EU Member States have been enabled to be more flexible in their approach, tailoring their response in a more individually suitable manner and the longer transitional period will help those countries, like the UK, with a heavy landfill dependence, there remain some areas of uncertainty and doubt as to the hard realities of implementation. The earlier Waste Framework Directive 75/442/EEC established the definition of 'waste': the Landfill Directive introduces references to 'municipal waste' using a broader definition of this category than is normal for the UK, and includes 'commercial, industrial and institutional and other waste which, because of its nature or composition, is similar to waste from households'. Throughout both the UK and Europe, less is known about these wastes than 'domestic' MSW in terms of their fractional categorisation and it was, therefore, not immediately clear how much of this would be biodegradable.

There was some initial suggestion that the targets might be applied to waste covered by the UK's more limited sense of the term only and for a while this remained an area of dispute. However, a House of Lords Select Committee report on the matter in March 1998 concluded that such an interpretation was not supported 'from a reading of the Directive as a whole'. Additionally, the legal opinion obtained in November 1997 by the UK's Environmental Agency (EA) that packaging waste from sources other than households should still be considered 'municipal waste' under the terms of the prior Incineration Directives, would seem to support their Lordships' position.

These earlier directives, which came into force in 1989, set much tighter operating standards, not least for pollution control. The aforementioned legal opinion effectively upheld the wider meaning of 'municipal waste' as defined by the Directives and thus much more material became subject to the tight emissions controls applicable to municipal waste incinerators. This promptly prohibited the burning of quantities of packaging waste in cement kilns, which had been proposed, since the stringent standards required could not be met.

The intended scope and meaning of 'biodegradable' is, likewise critical, since this clearly determines which wastes are covered by the legislation. The Directive specifies that such materials are 'capable of undergoing anaerobic or aerobic decomposition' and explicitly includes paper within this definition, which, as will be discussed in a later chapter, has some interesting implications for its best long-term use. Arguments of definition aside, by 2003, Member States must have implemented a strategy for the reduction of MSW-derived biowaste entering landfill, in line with the stages set out. However, should its generation actually increase above the 1995 levels taken as the baseline in the formulation of these targets, then, clearly, the relevant actual reductions would have to be greater than the stated percentage in each case. Since there is a general agreement that the amount of waste arising increases by around 1–3% per annum, this is by no means an unlikely scenario.

Finally, in shifting emphasis from a TOC-based approach to a defined target one, a potentially significant difficulty is created for those who seek to perform precisely the kind of biowaste pre-treatment that the Directive seeks to encourage. The products of AD and composting are themselves not fully biodegraded; if they were, they would have no use as soil additives. Even incinerator ash has, to some extent, a degradable content. Unless the Directive is more specific as to what constitutes 'biodegradable' waste, possibly by some reference to TOC, then there remains the possibility that any such material requiring disposal would count towards the allowed quota, thereby reducing the overall amount of untreated waste which could be accepted.

The Directive also heralds a ban on the co-disposal of hazardous wastes with municipal or other wastes, which must be implemented by 2004 on existing sites and will apply to all new landfills from 2001. From 2002 existing sites still to take hazardous waste will have to be classified as such and a total ban on the landfilling of liquid wastes at these facilities will be enforced. Liquids may be disposed of at existing non-hazardous sites until 2009.

After the full ban comes into effect in 2004, hazardous wastes will be consigned to dedicated monofills, excepting materials which have been treated by processes of solidification or vitrification to render them more inert, which may be mixed with non-hazardous waste provided their leaching qualities are similar and no biodegradable matter is involved.

There are a number of major implications of the Directive for existing landfill sites. Concerns regarding the methane generation from landfill sites were one of initial driving forces behind the European Union's original desire to act. Consequently, landfill gas must be collected and either used for energy generation or flared if this is not possible. Currently only around 160 sites of the 499 accepting 'significant amounts' of putrescible material do so. The UK produces around 27 million tonnes of MSW yearly, roughly one tonne per household, and typically around 30% of this is biodegradable. At present, only around 8% of UK MSW is incinerated; a further 8% is recycled and landfilling accounts for the rest.

To deal with the other major landfill pollution potential, all sites except those taking inert waste must employ leachate collection and comply with stated universal minimum liner specifications, instead of the UK's current risk assessment based approach.

By 2002, operators will have to have produced plans detailing how they will comply with the Directive requirements. The Environmental Agency, as regulatory body, will then have to make the decision as to which sites may continue and which they will cause to be closed down 'as soon as possible' and there is a requirement for 'adequate provisions, by way of financial security or any other equivalent' before operations can be carried on. This will apply to all new sites at once and to all existing sites by 2009.

The Directive states that all wastes must be 'subject to treatment' before consignment into a landfill, except material which is itself inert or for which such treatment would neither reduce its quantity nor its 'hazards to human health or the environment'. Originally this was intended to mirror French legislation, requiring some measure of separation or biological or chemical treatment before disposal. However, this has been substantially weakened during the course of negotiations to the extent that the definition of 'treatment' is now so widely drafted that it might potentially include the mere act of compaction within the collection vehicle, since it permits any treatment which alters the 'characteristics of waste in order to reduce its volume . . . facilitate its handling or enhance recovery'. Sorting of the waste prior to landfilling is expressly included under the Directive definition, which may serve to kick-start the subsequent biological processing of the biowaste fraction, as well as aiding the traditional recycling initiatives.

The Landfill Directive is the culmination of over nine years of European debate, with wide ranging implications for the whole waste industry, notably in those countries with a heavy traditional dependence on landfill. At its core is a phased reduction in the amount of biowaste destined for landfilling, though there remain areas of uncertainty over the definition and scope of its intended extent. Member States must devise their own mechanisms to ensure compliance and it seems highly likely that some form of financial instrument will help, at least in part, towards achieving this for most of them. Though beyond the scope of the current discussion, there are other aspects of the Directive which have equally important implications for waste management in general and a landfill lobby anxious to protect its own historical practices. In essence, the Directive represents the latest in the EU move towards the centralised harmonisation of waste practice and relevant environmental law across the national boundaries of its Member States.

The American Situation

In the United States, by contrast, the treatment of MSW-derived biowaste is less centrally regulated by the Federal government, which has largely confined itself to setting standards for the more potentially problematic sewage sludge. Hence, there are no national standards for other forms of biowaste and individual states are responsible for producing their own regulations, which leads, inevitably, to some variation across the country. There have been calls from certain quarters to extend the scope of the applicable sewage sludge regulations to cover MSW-derived biowaste operations, though in practice, any material or process giving rise to a pollutant potential approaching that of sewage sludge would already be required to comply with these limits. While it is fair to say that the treatment of the household biowaste fraction does not receive the same extensive scrutiny at national level, an awareness of the control on sewage sludge is essential to understanding the standards for biowaste in the US.

Issued in February 1993, the *Clean Water Act*, specifically the portion of it known as *Title 40 of the Code of Federal Regulations, Part 503 — The Standards for the Use or Disposal of Sewage Sludge*, established baseline limits for the United States. Arising from some twenty years of research, the '*Part 503 rule*' or simply '*Part 503*', as it is most commonly referred to, is a fundamentally different kind of regulation from what is usual in Europe, since it is based on risk assessment rather than the precautionary limits approach typically used elsewhere. This European model is sometimes termed the 'no net gain or degradation approach', meaning that there should be no net accumulation of pollutants in the soil nor any degradation of soil quality from pre-existing levels. The American regulation, however, examines the potential risk posed by selected key pollutants to humans, other animals and plants, making evaluations of a number of different possible pathways, from the direct 'single instant' scenario, to a lifetime of possible exposure via bioaccumulation routes. Thus, the standard ultimately set for a given contaminant will be based on the lowest concentration which was, according to this process, deemed to present an acceptable risk.

This approach has led to the acceptance of typically higher heavy metal concentrations and cumulative loading rates than would be permitted in Europe, since the ability of soil to lock up these metals indefinitely has demonstrated sufficiently well by extensive scientific research, over many years, to satisfy the American Environmental Protection Agency. Based on the scientific evidence, they are of the opinion that even if background levels of a particular heavy metal species were to increase slightly over time, its migration or availability for plant or animal uptake would be precluded by the combination of the resident micro-fauna and other general edaphic characteristics.

The Part 503 rule has been a major influence on many new state regulations on biowaste treatment, with the scope of its environmental consideration extending to ground water, surface run-off, soil and air quality. Though the limits thus established for sewage sludge do not directly apply to other forms of biowaste, the flavour of 503's requirements for contaminant monitoring, pathogen inactivation, odour control and site access restrictions can be seen in a number of states' recent biological waste treatment legislation. Additionally, the Rule established the idea of Exceptional Quality (EQ) sludge, which relates to materials which can be shown to be almost pathogen free and meet stringent pollution standards. Once the qualifying requirements have been achieved, the use and sale of such an Exceptional Quality sludge and, most crucially, products derived from it, are effectively unregulated. Achieving this standard is, then, highly desirable for the operators of treatment facilities. Again, though essentially a sewage sludge concept, the EQ ethos has also begun to enter the wider biowaste arena.

However, by virtue of the un-centralised nature of the approaches taken, the reality of biowaste treatment regulation is complex. Even within the scope of Federal limits, the resultant actual management and enforcement issues arising fall squarely within the remit of the individual states themselves, even to the extent of setting their own land application criteria, when necessary. For other forms of biowaste, the relevant legislation and requirements governing treatment are wholly state-based and there has been a considerable evolutionary change in both the form, and intent, of them over the years.

The so-called 'Garbage Crisis' of the mid-80s fuelled renewed interest in composting and other methods of biological waste treatment, since incineration lay virtually dormant amid public and regulatory fears over emissions and there was increasing concern over the apparent lack of availability of landfill void-space. At about the same time, responding to the same kind of concerns that ultimately led to the Landfill Directive in Europe, a number of individual states enacted legislation to exclude yard trimmings, food scraps and other biowaste from both landfills and incinerators. In 1988, there were fewer than a thousand biowaste treatment sites in the whole of the United States; a decade later, the number stands at 3,807[4].

Back in the late 80s and early 90s, biowaste legislation largely consisted of isolated, and typically short, passing references in the relevant state solid waste regulations to agricultural or yard waste composting. Thus, the Guidelines for Composting Yard Waste for Arkansas, which were issued in 1992, covered the ground in little more than a page. Drawn up a year earlier, Georgia's solid waste rules on composting also ran to slightly over a page, though conceding that more than just 'yard trash' could be usefully treated in this way. Michigan dealt with the matter in under a single sheet, though allowing a second one to expand the discussion to other potential uses of the process. In most state regulations during this period, the most prominent mention of composting was regarding its potential use as a Process to Further Reduce Pathogens (PFRP) in respect of the application of sewage sludge to land [5].

Throughout the 90s, more comprehensive regulations emerged often with stringent requirements for siting, operational procedure and after care provisions, principally aimed at the treatment of 'yard' and solid wastes. At roughly the same time, specific heavy metal limits and other standards appeared in some states requirements. Much of the legislation relates to the kind of biowaste to be processed, since the application of biological treatment in America has become increasingly diverse. Accordingly, while California, Illinois, Colorado, Maine and Massachusetts all address source-separated feedstocks, food waste, and other relatively clean biowastes, a number of other states have extended the scope of their regulations to cover additional named types, while others have brought all biowaste treatment under one unified code. Hence, from the initial stance on sewage sludge treatment, states have variously included abattoir, industrial process, dairy, fish, and grease trap wastes in their regulations.

Even so, the Part 503 sewage sludge ancestry can still be broadly seen in that salmonella and faecal coliform bacteria have been widely adopted as indicator pathogens, while a large number of states have imported the 503 heavy metal limits directly into their regulations.

However, in much the same way as the EU Landfill Directive is expected to drive European biological waste treatment, little growth in the number of MSW-derived biowaste processing facilities is thought likely in America, until more states impose their own landfill bans.

Product Standards

A number of existing biowaste treatment facilities, and many proposed new schemes, depend on some kind of open sale or other revenue generation to maintain viability. Consequently, before leaving the topic of regulation, a final mention should be made of product standards.

This has been a major area of contention for a number of years, since there are no truly recognised, universal standards for biowaste-derived products, with Member States of the EU and individual states of the US being free to set their own. In the public consciousness, heavy metal content is seen as the defining issue, and even people with little or no scientific training have some awareness of the perceived risk that they pose. Consequently, the lack of an established standard has made the acceptance of these products slow and uncertain. It has been widely acknowledged that such a system is essential to increasing the confidence of potential users in the performance and safety of such materials, while also promoting fair competition amongst producers. As it stands, the customer is presented with a bewildering array of 'standards' and an equally diverse approach to product labelling or description, as the following table serves to illustrate.

Table 3.2 Maximum Permitted Content of Potential Toxic Elements (mg/kg dry mass)

Element	Eco-Label	Organic Waste Composting Association
Arsenic	7	15
Cadmium	1.5	10
Chromium	140	1,000
Copper	75	400
Lead	140	250
Mercury	1	2
Molybdenum	2	10
Nickel	50	100
Selenium	1.5	5
Zinc	300	1,000

The EU Soil Improver Eco-Label includes biowaste-derived products, but is not exclusively concerned with them and was designed to enable the public to identify materials with the lowest life-cycle environmental impact. While the revised criteria shown here were an advance over the original 1994 version, which related very poorly to post-biologically treated products, there are still problems in applying this standard to MSW-derived composts and soil conditioners, most notably in terms of the permitted nitrogen.[6] The European Standards Committee CEN/TC/223 is working towards an appropriate system of standards, though this will again not be specific to biowaste-derived material, concentrating instead on the wider general categories of 'soil improvers' and 'growing media'. There is some further concern arising as a consequence, since the standards will relate purely to the final material and not to the means by which it arises. In the context of biological waste treatment, the method of processing, particularly in respect of assurances of product sanitisation, is a very important consideration in the building of consumer confidence. Clearly, an official standard or accreditation scheme which addressed these areas would represent a significant advantage to the marketing and acceptance of waste-derived composts and other soil products.

It must be remembered, however, that such products may be used in a number of ways and in a number of different circumstances. Thus they can form mulches, growing media or soil improvers, for applications in agriculture, horticulture, amenity landscaping or land restoration. There would seem to be a clear case to be made for a range of 'use-based' standards to take account of this, enabling both users to select the most appropriate product for their needs and producers to have a clear quality benchmark to achieve. Some states of the US have moved a little way towards this, with new legislation which eases biowaste soil products into the existing regulatory framework for soil amendments and fertilizers, for example requiring NPK analysis, as in Minnesota, New York and Maine. There remains some way to go, though, before consumers, either in the US or the EU can hope to be presented with a readily understandable system and its continued absence acts as a significant limit on the wider acceptance of biowaste-derived materials.

The regulatory framework as it applies to the processing of biowaste is, then, like biological treatment itself, rather in its infancy in terms of the legislature. Existing law provides the overlying structure, with new rules emerging to meet the needs of a changing perception of both the nature of waste and the purpose of waste management. Like any legal interregnum, there are anomalies and areas in which either the legislation appears to conflict with its intended purpose, or its exact interpretation has not yet been fully explored. It must be borne in mind, however, that this represents merely a stage in the evolution of functional regulations to govern the growing application of biological techniques to the matter of biowaste. In both the US and the EU, the realisation has dawned that effective diversion and recycling strategies can only be achieved by addressing precisely this issue. Some of the reasons for this will be discussed more fully in the next chapter.

References

1. *State Greening Team Targets Organics Diversion*, BioCycle Magazine, August, 1998, p. 44.
2. The *State of Garbage in America* survey, BioCycle Magazine, April, 1999, p. 66.
3. *Part 243 Guidelines for the Storage and Collection of Residential, Commercial and Institutional Solid Waste*, Subpart B – *Requirements and Recommended Procedures*, Section 243.200–1 Paragraph (a) *Storage requirement*.
4. The *State of Garbage in America* survey, BioCycle Magazine, April, 1999, p. 64.
5. David Riggle, personal correspondence.
6. *Report of the Composting Development Group on the Development and Expansion of Markets for Compost*, Department of the Environment, Transport and the Regions, July 1998, section 2.3, p. 11.

CHAPTER 4
Biological Waste Treatment

The objectives of treating biowaste by biological means are very simple and can be summarised as the reduction in the waste's potential to adversely affect health or the environment, the reclamation of valuable substances present and the generation of a useful final product. In the widest sense, this means the conversion of the biodegradable matter by micro-organisms to yield a bulk-reduced, stabilised residue, the original complex organic molecules having been broken down into chemically simpler substances, to facilitate their recirculation in a biological context. The relative importance of these aims may be viewed as hierarchical in nature, as shown below.

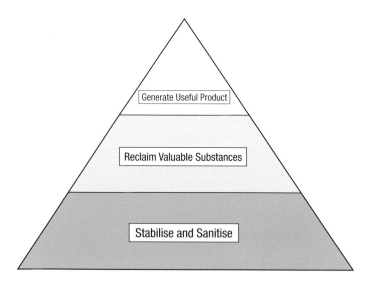

Figure 4.1. The pyramid of priorities

Removing the risk of environmental or public health damage and deriving a stabilised product must always form the lowest step in the hierarchy, since this is the first aim of all biological waste treatment. Whatever the material is ultimately to be used for, it must be rendered safe in human and ecological terms and part of achieving this, of course, requires that it is effectively made biologically inert. The next level up involves the reclamation of substances from within the biowaste which have some value or potential for reuse, which include carbon, nitrogen, phosphorus and potassium, together with other chemicals and trace elements which have a role in living systems. In many respects, this is closely linked to the stabilisation process, precisely because of this aspect of the nature of these chemicals, which, if left within the untreated material, would always represent the potential for further microbial activity. At the apex of the pyramid, the making of a useful end-product is itself largely linked to the reclamation and stabilisation requirements of the preceding two levels and, clearly, the final use depends on the efficacy of those steps. Hence, the levels within the pyramid, though shown as distinct for clarity, should more properly be viewed as gradation of effect and the priorities assumed to form a spectrum, rather than being strictly hard-edged.

The practical outcome of the application of these principles results in a number of environmental advantages:

Reduction of Biowaste Volume Destined for Landfill
- Reduction of uncontrolled gaseous emissions to atmospheres from landfill sites
- Reduction in the total amount of greenhouse gases
- Void space freed up for materials for which this route is the most appropriate

Generation of Soil Amendment Product
- Possible reduction in demand for peat
- Possible reduction in use of artificial fertilizers
- Potential to improve soil quality and fertility
- Potential to reduce erosion

Stabilisation and sanitisation are central to the whole practice of biological waste treatment, both in respect of processing and of product marketability. Again, to achieve the goal of a useable end-product, the basics must be right. Stability of the final material is the key to ensuring that a consistent and quality product is brought to market; that product's assured freedom from weeds, pests and pathogens is what engenders customer confidence. Though the necessity of each is readily understood, the determination of both is worthy of further consideration.

In the public perception, a stable product would be defined by largely subjective criteria of appearance, feel, or smell. These are difficult to standardise and impossible to measure, but it is clearly important to bear in mind the likely layman's view, to whom a full chemical assay may be of much more limited value. In practical terms, stabilisation may be defined as adequate biodegradation to ensure that the material can be stored normally, even in wet conditions, without giving rise to any problems. Under similar circumstances, a biowaste-derived product which has not yet achieved proper stabilisation typically might begin to smell, show signs of renewed microorganism activity, or attract flies or vermin, for example. Direct respirometry of the specific oxygen uptake rate (SOUR), as a window on microbial activity within the post-treated matter, is gaining support as a potentially uniform, quantifiable measure of stabilisation. There is, however, still a need for some caution in interpreting the figures, at least until the method becomes more widespread and better baseline data have been obtained.

Sanitisation is often wrongly seen as synonymous with sterilisation. The confusion arises because of the universal awareness regarding the necessity for biowaste-derived products to be free of pathogens. Clearly, this is an important consideration and forms a major part of the biological waste treatment pyramid's foundation, being one half of the strategy for the removal of potential risks to health, the other half being the exclusion of directly toxic substances. However, the goal of sanitisation is not to produce a biological wasteland, devoid of all life, since one of the benefits to poor quality soils which can be provided by a good soil enhancer is the ready-made, thriving microbial community it brings along with it. The need to remove the potentially harmful organisms, seeds and spores, both for reasons of public health and final product quality, while not compromising the activities of beneficial microbes, is one of the required balancing acts of biological waste treatment and one for which the precise details depend on the technology being used. The whole issue of pathogens, sanitisation and the public perception of these refuse-derived products has significant repercussions for the development of this waste management sector and it is one to which we shall return to consider in greater depth, later in this chapter.

The scope for biological treatment of appropriate waste is governed by a number of factors, though the individual contribution of each may vary. Principal amongst these is the true availability of the biowaste, and this naturally falls into two further parts. Firstly, the actual type, amount and suitability of the material itself and secondly, the manner in which it is collected and the physical condition in which it can be presented for treatment. Clearly, there is an abundant supply of suitable material. The UK produces about 27 million tonnes of MSW yearly, of which something around a third, roughly 9 million tonnes, is biowaste. As we saw in chapter 2, it has been estimated[1] that only 3.4% of that is recovered and treated, this figure including home composting. In the European Union as a whole, the rate is around 15% of the estimated 60 million tonnes or so of recoverable biowaste[2] being generated annually. The relative contribution of each Member State to this approximate total of 9 million tonnes subject to biological treatment within the EU is shown in Fig. 4.2.

It is important to view this larger picture, rather than relying on the rates for individual countries, which can be misleading in the wider context, since there are four major producers of biowaste in the Union, namely France, Germany, Italy and the UK. While Germany, for example, manages around 45% both domestically and as its contribution to the EU total, Holland's higher home rate of 90%, though laudable, is on a smaller tonnage, hence representing only 20% on the European stage. For the same reason, Austria's 50%, becomes 12.5% and so on. Interestingly, of the remaining three major producers, although their percentage rates are very much lower than some of the other countries, all manage to better their home figures in this kind of comparison. Thus, France makes a contribution of 4.5% to the European total as against 3% at home, Italy rises 0.3% to a European 2.3% and the UK's European share is 3.6%, 0.2% higher than its domestic rate.

What is perhaps more interesting still is both to consider these figures against the backdrop of current biological waste treatment and in terms of their implications for the stated intentions of the European Parliament, in respect of the future of biowaste management. The Landfill Directive will, ultimately, cause the exclusion of a huge mass of biowaste, even if it takes a little time to bring this about. The initial

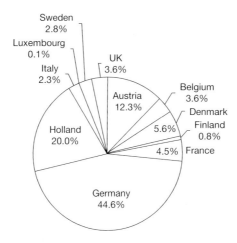

Figure 4.2. Member States' contribution as a percentage of the EU total biowaste treated.

25% reduction by 2006, for those countries which do not make use of the extra time provisions to comply, will probably make little immediately noticeable impact across a Europe already averaging 15% biowaste recovery. By 2020, however, when all Member States, even those which opted for derogation, must achieve a 65% diversion target from their 1995 levels, this amounts to nearly 40 million tonnes of biowaste which will require alternative treatment throughout the EU. This represents close on four and a half times as much as is currently accomplished. Even Germany, which makes the highest individual contribution, at the present time manages only 6.6% of the total available biowaste arising within the Union and the other countries rather less, as shown in the following graph.

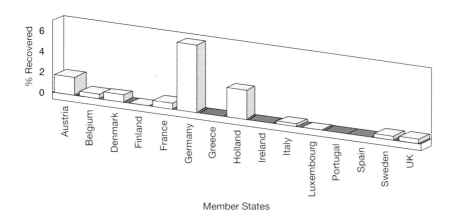

Figure 4.3. Percentage of total EU available biowaste recovered by each of the member states.

Germany began the introduction of separate collections for biowaste in 1985, at which time there was almost no centralised treatment capability in the country. This situation had changed to allow the processing of around 500,000 tonnes per annum five years later and the total has continued to rise over the intervening decade to the current level.

Much of the initial success of biological treatment has concentrated on the so-called 'green' or 'yard' wastes, the plant matter from domestic, commercial and municipal gardens. Since such material is already effectively segregated at the point of origin, generally highly biodegradable and, in many local authority areas, already demands a disposal route other than simple consignment into the dustbin, there is an easy fit to the requirements of alternative management practices. In the UK alone, the production of this type of biowaste is thought to run to around 5 million tonnes annually[3]. Clearly, this is one particular area in which biological waste treatment can readily make rapid advances. In America, for example, the processing of yard waste has increased greatly throughout the 90s, currently achieving better than 40% recovery which represents a contribution approaching a quarter of the overall US national recycling rate.

Food waste is often thought of as less easy to integrate into a biological treatment regime since to present it for processing generally requires either a greater degree of work by the householder or an efficient separation system at the treatment plant. Moreover, it typically contains meat and fish scraps which can bring potential vermin problems. While the Dutch mandatory separate collection of biowaste has led to a 90% recovery rate, in the US, 41% of food waste is destined for landfill and 37% ends up at the local sewage works, via kitchen sink grinders. While in one sense this simply shifts the problem to somewhere else, there are a number of points which lend support to this approach. Not only does it deal with a significant amount of biowaste which would otherwise be entering landfills, it does so in a way which reduces odour, fly and vermin nuisance, dove-tails well into the existing biological treatment capabilities and, by acting as an additional source of carbon to augment the already nitrogen-rich sewage, benefits both process and product. One recent piece of work[4] has gone as far as to suggest that this route may, under certain circumstances, actually represent the Best Practicable Environmental Option (BPEO). However, there are, equally, factors which indicate against its widespread adoption, ranging from the cost of the grinder appliances and concerns over sewerage carrying capacity, to the implications for water demand and the ability of current treatment facilities to take up the extra load.

To some extent, the discussion of biowaste types suitable for biological treatment is irrelevant, since the European Landfill Directive, local bans implemented by individual American states and any likely future laws enacted, anywhere, expressly to address the issue of biowaste entering landfills, apply, of necessity, to all forms of such material. To attempt to do otherwise would be a practical impossibility and an enforcement nightmare. While some biowaste types lend themselves more readily to processing, the wider scope of biological waste treatment is not limited solely to them. This leads naturally on to the second of the facets of availability, namely the manner in which the material is collected and the physical condition in which it is presented.

Collection Regimes

Waste arrives for treatment as a result of one of three general means of collection; as mixed MSW, as part of a 'separate collection' scheme or via civic amenity sites or recycling banks. Although any of these methods may be applied to all household waste, each having its own particular advantages and disadvantages, viewed simply from the biowaste perspective, they may be briefly characterised in the following manner:

Mixed MSW

This traditional collection of unsorted refuse, generally destined for landfill or incineration, is not ideal for the recovery of any kind of material from the waste stream and, in particular, it requires considerable additional work to provide a suitable feedstock for processing by biological means. The risk of cross-contamination is, obviously, the highest of all for this method, and any attempt to effect biological treatment on mixed waste collected in this manner is a clear non-starter.

Separate Collection Schemes

These vary greatly, dependent on the required emphasis of the local waste initiatives in place. Many such schemes are based around the commingled dry recyclable concept, requiring the segregation of waste into two broad fractions, 'dry recyclables' (paper, glass, cans, plastics etc.) and 'the rest', which are stored and collected separately. There are two approaches to this in general use. One is to have two rubbish containers, the disposables being picked up weekly in the normal manner, with a fortnightly extra collection, most commonly using a different type of vehicle, for the recyclables. The other method typically uses one bin, split into two compartments, which is serviced in a single visit. This latter system avoids the additional cost of putting on extra men and waggons, which is one of the major disadvantages of separate collections. There are a number of sophisticated designs of vehicles which get around the potential labour-intensification of such collections by mechanically lifting the householders' wheeled bins against internal baffles, set to coincide with the divisions on the bins themselves.

Though the standard 'commingled' route leads to a biowaste-rich, but essentially contaminated fraction, the idea of separate collection can be successfully applied to yield an excellent feedstock for biological processing, as a number of countries currently do. It is not, however, without some drawbacks, which will be examined in more depth, later.

Recycling Banks and Amenity Sites

The specific designation of the receptacles at these facilities again generally depends on local emphasis, being categorised very specifically, for example, into different waste types or more loosely into much larger, general groups, with the intended final management option largely deciding the level of segregation required. It is widely held that the 'green waste' (i.e. garden/yard waste) fraction from these sites is the cleanest source available for centralised biological treatment and waste of this type accounted for almost three-quarters of the biowaste processed in the UK[5], in 1997.

It should be readily apparent that the method of collection is the single most influential factor which dictates the physical condition of the material delivered to the treatment plant. This, in turn, has clear implications for the required steps between waste arrival and processing, which cause the biowaste to be rendered into a suitable form for the particular biological treatment which is to be employed. The degree of cleanliness required will vary between technologies and with operational practice, but it is a general, and somewhat obvious, rule of thumb that the more segregation that is done at source, the less equipment, time and cost will be demanded on-site. Human error aside, given the current state of 'dirty MRF' technology, the resultant feedstock from a separate collection will always be more readily treated than that from any form of post-producer sorting. This is not to say that there is no future in either MRF design or mixed waste treatment, but it is inevitably the case that the options for a biowaste fraction thus derived are more limited.

The method of on-site sorting used needs to be matched to both the incoming waste stream and the available local resources. For facilities taking commingled dry recyclables, in areas where there is a premium on the provision of employment, then a heavy reliance on hand picking is a viable option. When the desired separation target is biowaste, however, it is not realistic to look to use people in anything other than a supervisory capacity, particularly if the waste collection system is a fully mixed one. There is an underlying difference of ethos, dependent on whether the driving force behind the need for sorting is to produce a suitable material for biological processing or to maximise recycling *per se*, across the board. The latter case has much in common with the dry recyclables approach, seeking to derive a high value product in the form of its separated fractions. However, it is difficult to see how this goal can ever be entirely compatible with taking a mixed MSW feedstock. Despite largely unsubstantiated claims from certain quarters to the contrary, there seems little likelihood of any commercially credible, mechanical separation system arising in the foreseeable future, which can effectively reclaim materials from such a source. Aside of the technical difficulties implicit in achieving the required levels of purity and cleanliness by secondary sorting, the financial considerations alone would make what is recovered far from cost-effective to return into the established recycling market.

There is more purpose to be served in further developing those systems which facilitate the production of rather broader categories of waste from mixed MSW, an approach which enables the bulk of the biowaste to be removed. Though nowhere near as pure a fraction as source-separated green waste, this does, first and foremost, represent a valid diversion from landfill or incineration routes, while also opening the way for some kind of biological treatment, particularly if the worst of the non-organic inclusions can be removed without too much effort or outlay. The remaining material types obtained by such systems, currently unusable in their contaminated form, can at least be disposed of under existing arrangements, pending the establishment of new outlets which may find these products acceptable. This is, of course, more akin to the idea of separation as a means of feedstock preparation for bio-treatment. In many ways, this will become the major justification for mixed waste MRFs in the future, since they are an uneconomic means of recovering recyclable materials, but viewed from a biowaste standpoint, an excellent method of landfill diversion. Moreover, as the volume of mixed MSW requiring biowaste removal is set to grow, it seems probable that the complexity of the separation systems will reduce, as this simple reality of their true future contribution becomes apparent. With a shift in emphasis to organic extraction rather than high-value material recovery, those countries which retain a mixed MSW collection may well see an increase in rotary drums, pulverisers and the like used to achieve gross biowaste removal. There will be little need for the finesse of air classifiers, electro-magnets and eddy-current separation, though these devices will continue to do sterling service on clean and separately collected, dry recyclable streams. In the final analysis, perhaps the wheel will turn full circle and the whole business of post-producer sorting disappear. How many authorities ultimately abandon mixed household collection in favour of mandatory source segregation remains to be seen. Which approach makes more economic sense largely seems to depend on how the financial model is loaded and what additional factoring is applied to job creation, capital expenditure and labour costs.

Having obtained a separate waste fraction rich in biodegradable material by whatever means, the next important consideration is its physical form. This is of far more fundamental importance to biowaste than to any other material reclaimed from refuse. For plastic, aluminium, steel or paper, the question of whether it must be delivered baled, shredded, chipped, crushed or loose is simply one of convenience. For waste material destined to undergo biological treatment, the issues of size, consistency and relative purity are indispensable elements of the process, since their optimal values are directly defined by the needs of the micro-organisms responsible.

Generally, all large items of biowaste will require to be broken into smaller particles for processing and most biological treatment facilities use some form of shredder or pulveriser to produce the required size. At the macroscopic level this makes mixing and homogenisation easier; at the microscopic, it raises the surface area to volume ratio, thereby increasing the feedstock availability to microbial action. Since the object of biological treatment is the controlled breakdown of the original complex organic molecules, this mechanical intervention both contributes to this conversion on the gross scale, while rupturing the structural integrity of tissues to enable the same process to begin at the cellular level.

There has been much speculation over the years as to the 'ideal' input for biological treatment and the best means to produce it. In absolute terms, this is an impossibility, since any material, outside of the universally applicable requirements for biodegradability and relative purity, can only be the ideal for a given processing regime. With feedstock and treatment, of necessity, so intimately bound together, the inherent variability of waste arisings can itself represent a limit on process performance, such that material and system may more truly be regarded as the optimum compromise. As the following diagram illustrates, while the background level of food waste arising remains broadly constant, the overall biowaste production rate increases markedly over the summer months, especially in temperate zones.

It has been estimated that, in some authority areas, around 80% of the garden refuse contribution to the overall biowaste total is made over only 22 weeks of the year[6], though this is obviously very dependent on a variety of local factors. Consequently, any biological treatment initiative needs to take this seasonal influx into account when designing plant throughput capacity. Equally, staffing levels and any transport or storage provision also need to be planned with this in mind. This can be of particular relevance to tourist areas, which experience the double-whammy

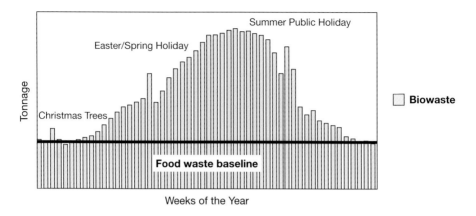

Figure 4.4. Illustrative graph of biowaste in MSW over the year.

effect of both this normal increase in gardening activity, coupled with the additional upsurge of food waste generated by the temporarily resident extra population throughout the spring and summer months. Strictly speaking, the loading on the biowaste facilities of holiday areas also rises by virtue of the supplementary demand for sewage treatment, though this is outside the scope of the present discussion.

There has been a considerable amount of work done on the variability of household waste over the last ten years and the effects of locality and socio-demographic factors are now widely appreciated. However, from the point of view of the production of a feedstock suitable for biological processing, the variations over time of the relevant proportions of potential contaminants may be more important. This group includes not only the non-, or slowly, biodegradable materials, like paper, plastic or brick rubble, which simply get in the way and take up processing space, but also the actively toxic or inhibitory. At its simplest, the largest amounts of biowaste are being generated within household refuse at exactly the same time as there is the greatest likelihood of significant quantities of pesticides, herbicides and fungicides being delivered to the treatment plant. In one respect, there is the argument of dilution, but in the end, it comes down to a question of concentration. This in turn, leads on to the issue of product quality, which is one of the major factors in biowaste processing today.

Quality is the key to achieving the final requirement for effective biological waste treatment, the generation of a useful product, which is itself critical in helping to decide the processed material's ultimate fate. Moreover, it is important to consider how this affects the end-use, not only in absolute terms of the product's chemical suitability, but also in respect of how the perception of quality may influence it. This area of product perception is possibly the easiest to understand, and yet the hardest to rectify, of all the many considerations affecting biowaste. As discussed in the previous chapter, the lack of an agreed standard for biowaste-derived materials is a considerable obstacle to the adoption of these soil products by both industry and individuals. Their wider acceptance for a number of applications for which they are ideally suited, is hampered by concerns of quality and consistency, and without a meaningful and common yardstick against which the various competing products can be measured, this resistance will remain. The potential market is enormous, and it will need to be, as the levels of diversion from landfill increase over the coming years. However, even the most forward-thinking of landscape gardeners, farmers and householders will be reluctant to commit to using a material without some sort of assurance as to its likely effect which they can relate to products they already know and have previously used. There is a willingness to embrace the new, as a recent telephone survey[7] of professional landscapers found. All of those asked said that they would consider a waste-derived compost or mulch, but the major non-economic barriers seem to be worries over quality and consistency coupled with the lack of established performance data or growth trials. Availability was also cited as a difficulty, but this is something of a chicken-and-egg argument, which the more generalised move to biological waste treatment for compliance, rather than commercial, purposes seems likely to force out of the current impasse. The survey's sample size was small, but the message is clear; every single one of them would consider biowaste-derived materials, if only the way was made easy for them to do so. It is not simply that these professionals themselves must be convinced as to the safety and benefits of the products, these being two distinct aspects; they must also be able to sell the idea to their clients and this inevitably means with some kind of guarantee. To expect otherwise is to ask them to gamble with their reputations and continued business.

Marketing Issues

The future of biological waste treatment is inextricably linked with the establishment of good and sustainable markets for the products, particularly as the physical amount of these materials being generated increases. This in itself will not automatically make the commercial viability of biowaste processing plants certain, but it will enable these treatment technologies to compete more directly with other waste management options. Moreover, it may mark a move away from the two extremes of product outlet which currently appear to be the only generally available routes, material being moved either at enormous commercial premium or effectively given away at little or no cost. Arguments abound in the biowaste industry as to whether it is realistic to seek to make money on the sale of derived soil improvers. It is difficult to make any definitive statement on the matter, since the economics of individual plants depends so greatly on factors specific to each and its locality. However, those projections of compost revenue based exclusively on the garden centre market, where some of the specialist proprietary materials sold in small bags may fetch anything up to the equivalent of £150 a tonne, look increasingly suspect. Equally, the market for poorly produced, heavily contaminated, so-called 'compost' as a day-cover for landfills must also be of limited duration, even if the terms of the Landfill Directive would continue to permit the use of such lightly-processed material in this way. However, this avenue will remain a valid outlet, perhaps for biowaste which has been more fully treated and properly matured, which cannot meet the requirements of other markets. Hence, even the appropriate fraction of mixed MSW collections, secondarily separated from refuse, and subjected to a suitable bio-processing regime, can play its part in reducing the biologically active elements entering landfill, even though the glass, plastic and paper residuals would make such a material unacceptable for other uses. In an effort to increase their output, a number of incinerators are also investigating this approach, to divert non-combustible material from their grates.

If biowaste processing is to be successful in the wider context, addressing the concerns of potential users will be an essential step in the removal of the present barriers to acceptability. Undoubtedly, much of the general public's perception of MSW-derived soil products will respond positively to a good education programme, but it will be important to deal with these worries constructively and in a genuine spirit of openness. From the perspective of developing the market, quality control, product management and performance criteria issues are likely to pose less of an impediment than those perceived to be of a health, safety or environmental nature. The possible presence of heavy metals, pathogens and toxic, polluting or potentially injurious contaminants within the final product are the main areas which provoke most general disquiet. Consequently, it is likely that any proposed industry standards, as well as setting out the necessary final quality of the materials themselves, will also need to consider the processes involved, especially in respect of establishing minimum sanitisation and monitoring requirements. This is particularly relevant to any attempt to realise the potentially huge opportunity represented by agricultural applications, since today's agri-business is almost entirely driven by the needs of the supermarkets. The British Retail Consortium's action, which helped bring about the matrix safety code for the treatment and application of sewage sludge, stands as a clear precedent for the involvement of such major chains in the formulation of standards for soil products. The upsurge of 'organic' farming has already led to a proliferation of compost acceptability criteria throughout the world, though as has been discussed earlier, the existence of a series of differing benchmarks does not greatly benefit the average product user.

Heavy Metals

The potential for heavy metal contamination in waste-derived soil materials is a well known and frequently raised objection to the wider use of such products. Again, part of the problem is a created one, in that with so many individual standards defining what is acceptable, the absence of uniform criteria based on a scientifically sound consensus to provide a clear statement of what truly represents safe levels, leaves room for doubt (table 4.1). It is difficult to decide who, or more importantly, what, is right, when so many views are expressed.

Time and time again, recent history has shown that where new or unusual technologies are concerned, people are very reluctant to gamble with their health and safety. Risk perception is a curious thing, when viewed in absolute terms, odds of 1-in-14 million being thought high enough to sell countless lottery tickets, though the 1-in-7 million chance of being struck by lightning is discounted as 'unlikely'. But while the dangers inherent in driving a car, for instance, may be ignored since the activity itself is seen as a necessary, or even enjoyable, thing, 'imposed' risk seldom enjoys the same latitude.

The over-riding contribution made by the initial feedstock to the final product quality is a recurrent theme of biological waste treatment. This is as true in respect

Table 4.1 Some of the Various Permitted Metal Levels in Biowaste Products[8] (All in mg/l)

	Cadmium	Chromium	Copper	Lead	Mercury	Nickel	Zinc
Austria							
Class 1	0.7	70	70	70	0.07	42	210
Class 2	1.0	70	100	150	1.0	60	400
Belgium							
(Flanders)	1.5	70	90	120	1.0	20	300
Finland	3.0	–	600	150	2.0	100	1,500
France							
NF Compost Urbain	8.0	200	–	800	8.0	–	–
Germany							
RAL GZ 251	1.5	100	100	150	1.0	50	400
Holland							
Netherlands compost	1.0	50	60	100	0.3	20	200
Switzerland							
Osubust	3.0	150	150	150	3.0	50	500
UK							
Sewage Sludge Code of Practice Soil Limit	3.0	200	135	300	1.0	75	300
Eco Label	1.5	140	75	140	1.0	50	300

of heavy metal levels as for any other aspect, as the following data obtained from large-scale German composting operations reflects (table 4.2). In an attempt to put the matter of heavy metals into context, these levels are shown against the UK's Interdepartmental Committee on the Redevelopment of Contaminated Land (ICRCL) trigger concentrations for the selected contaminant substances. These were originally produced to assist in the selection of appropriate further uses for such sites and to determine the need for any remedial action. However, they are useful here to give an indication of the wider view of what had tentatively been suggested as an acceptable metal concentration in the environment, while also demonstrating the idea of what constitutes a permissible level being defined in terms of the intended purpose to which the material is to be put. Given the variance of the initial feedstock, its collection and processing methods, coupled with the widely differing potential applications for waste-derived soil products, this is an approach which the regulators of the biowaste industry would be sensible to bear in mind. Cost and common sense both mitigate against forcing the same standards on compost destined for landfill cover or motorway embankments as would be expected of a material to be sold to the general public for their gardens.

While the image of biowaste-derived compost in some minds is one laden with heavy metals, the reality, certainly for source separated material, simply does not bear this out as table 4.3 illustrates. When the same German compost levels from the

Table 4.2 *Illustrative German Compost Metal Levels[9] Shown with the UK ICRCL 'Trigger Concentrations' for Comparison (all in mg/kg)*

Metal	Mixed MSW	Source Separated	ICRCL Domestic gardens, allotments	ICRCL Parks, playing fields, open spaces
Cadmium	5.5	0.7	3	15
Chromium	71	34	600	1,000
Copper	274	50	130*	130*
Lead	513	68	500	2,000
Mercury	2.4	0.2	1	20
Nickel	45	21	70*	70*
Zinc	1,570	222	300*	300*

* The ICRCL define these as 'Group B Contaminants', which though phytotoxic, do not normally represent a health hazard. The concentrations shown relate to 'any uses where plants are to be grown'. It should also be remembered that they relate to a soil pH of around 6.5; if the value falls, both the uptake and the toxicity increases. Moreover, it is important also to appreciate that grasses are generally more tolerant of phytotoxins than other plants and their growth may not exhibit any adverse effects at or about these threshold levels.

preceding table (table 4.2) are judged against the national standards shown in table 4.1, it becomes clear that the material derived from the source separated biowaste meets almost all of these requirements. It only fails to comply with the most stringent standards for three out of the seven metals and of these, one failure is by just 1mg/l, which represents 5% over the limit figure. Source separated feedstock entering anaerobic digestion rather than a composting process can give rise to equally as 'clean' a final product, a fact often missed, as the comparison data from the Valorga plant at Tilburg shows, failing only the most exacting standard for one metal. As one might reasonably expect, the material coming from a mixed MSW feed fared notably less well, after both AD and composting, but again, not as badly as its reputation might suggest. On this basis, it seems that there is good reason to promote high rates of application for maximum conditioning effect as a genuine reality, at least for appropriately specified, source segregated biowaste soil enhancers[10].

With effectively ten standards (counting the Austrian ones separately) and seven metals, the scores for each material out of the possible 70 are:

    Source separated AD: 69
    Source separated compost: 64
    Mixed MSW AD: 27
    Mixed MSW compost: 19

The ranking itself is unfair, since the German figures are generic ones, while the AD results relate to two specific plants and it would be possible to find composting facilities which could better the quoted Valorga outputs. However, what is interesting is the close similarity of overall result for the two feedstocks, proving once again the central importance of input material quality.

The heavy metal issue is, clearly, an area where a sensible education programme to raise public awareness of the facts, rather than the received wisdom, about such materials will be an essential tool in the ongoing development of the product market.

Table 4.3 A Comparison of the Compliance with the Metal Content Standards (as previously shown in Table 4.1) of Soil Products derived from Source Separated and Mixed Wastes, by Anaerobic and Aerobic Treatment Technologies (all in mg/l)

	Cadmium	Chromium	Copper	Lead	Mercury	Nickel	Zinc
COMPOST Separated[a]	0.7	34	50	68	0.2	21	222
Compliance with table 4.1 standards	PASSES ALL	PASSES ALL	PASSES ALL	PASSES ALL	Passes all except Austrian Class I	Fails Dutch & Belgian by 1 mg/l	Passes all except Dutch & Austrian Class I
COMPOST Mixed MSW[a]	5.5	71	274	513	2.4	45	1,570
Compliance with table 4.1 standards	Fails all except French	Passes all but Dutch, Austrian and Belgian (last two by only 1 mg/l)	Fails all except Finnish	Fails all except French	Fails all except French & Swiss	Passes all but Austrian Class I Dutch & Belgian	FAILS ALL
ANAEROBIC DIGESTION Separated[b]	0.5	23	27	67	0.1	7.0	190
Compliance with table 4.1 standards	PASSES ALL	PASSES ALL	PASSES ALL	PASSES ALL	Passes all except Austrian Class I	PASSES ALL	PASSES ALL
ANAEROBIC DIGESTION Mixed MSW[c]	2	150	80	600	2	40	575
Compliance with table 4.1 standards	Fails all but UK, Finnish, French & Swiss	Fails all except UK French & Swiss	Passes all but Austrian Class I Dutch & EcoLabel	Fails all except French	Fails all except French, Swiss & Finnish	Passes all but Belgian & Dutch	Fails all except Finnish

a Previous German Data
b Valorga Plant at Tilburg, Holland[11]
c Valorga Plant at Amiens, France[12]

Health Risks

Another widely perceived problem affecting the treatment of biowastes is the potential health risk posed. This naturally divides into three main areas:

- The occupational health matters facing workers at biowaste plants
- The public health issues liable to affect those who live or work near such facilities
- The product-related concerns relating to those who use the end material or consume produce grown in it.

In all of these cases, the risk of disease comes from two main sources. Firstly primary pathogens, which are the disease causing agents contained within the arriving waste, their presence being largely dependent on the nature of the feedstock itself. Though this is a larger problem for sewage or animal manure treatment facilities, since raw faeces are not commonly encountered in MSW, pet droppings, disposable nappies, putrefying food and other routes do represent a contribution to the potential health risk to employees, which should not be ignored. Primary pathogens are generally bacteria, protozoa, platyhelminths or viruses and represent a potential risk of infection to healthy individuals. Secondary pathogens are those which may develop during the processing itself, typically in the case of centralised composting operations, fungi such as *Aspergillus fumigatus* and species of the genus *Micromonospora*. Secondary

Table 4.4 Common Primary Pathogens and Parasites: Thermal Inactivation (Adapted from World Health Organisation figures)

Pathogen / Parasite	Thermal Effect
Brucella abortus *Brucella suis*	61°C – death occurs within 3 minutes
Corynebacterium diphtheriae	55°C – death occurs within 45 minutes
Entamoeba histolytica cysts	68°C – death occurs
Escherichia coli	55°C – most dead within 1 hour 60°C – most dead within 20 minutes
Legionella	46°C – growth ceases 60°C – death in minutes 75°C – no viable *Legionella*
Mycobacterium tuberculosis hominis	66°C – death occurs within 20 minutes 67°C – death is instantaneous
Salmonella typhosa	46°C – growth ceases 55 – 60°C – death occurs within 30 minutes
Salmonella (others)	55°C – death occurs within 1 hour 60°C – death occurs within 20 minutes
Taenia saginata	71°C – death occurs within 5 minutes
Trichinella spirahs larvae	50°C – significant infectivity reduction after 1 hour. 62 – 72°C – death occurs

pathogens pose a threat to healthy workers by chronic exposure or to others with immune systems which are impaired, as in existing long-term illness or other prior infection.

As discussed at the beginning of the chapter, the requirements of product sanitisation are not the same as sterilisation, the goal being to make the final material safe to handle, while ideally retaining its beneficial microbial compliment to enhance the soil ecology of the soils to which it is applied. It is well known that many disease causing organisms are inactivated or destroyed by heat and, accordingly, much of the currently routine sanitisation practice, particularly at composting plants, relies primarily on this for its effectiveness. This approach has a logical appeal since it is widely appreciated that during the composting cycle, relatively high temperatures are generated, usually in the region of 55–65°C, and that these commonly persist for some time. Moreover, peak temperatures often exceed this range for short periods. As will be seen in table 4.4, most of the pathogenic micro-organisms are effectively inactivated by temperatures above 50°C.

In effect, then, composting tends to be viewed as largely self-treating, while AD, typically being run at lower temperature and relying on the lengthy period of oxygen deprivation to inactivate a proportion of the pathogenic bacteria, protozoa, persistent platyhelminth ova and viruses, is generally thought of as incompletely sanitising. As a consequence, there has been a widespread general trend to define sanitisation, at least for composting operations, in terms of the duration of exposure to a given temperature and some of these requirements are shown in the next table.

Table 4.5 Selected EU Member States' Compost Sanitisation Requirements[13]

Member State	Temperature (°C)	Duration in days
Austria	65	6
Belgium	60	4
Denmark	55	14
France	60	4
Holland	55	2
Italy	55	3

However, despite the widespread acceptance of this as a viable sanitisation route, it does require a certain degree of diligent management to ensure that the entire batch is suitably treated. While the principle of thermal inactivation of pathogens and parasites is well established, in practice, and specifically in the context of a mass composting operation, simply measuring a core temperature adequate for sanitisation cannot guarantee the product quality, when the outer surface may be at little more than ambient. It has been suggested, since this process is not an exact science, that material being composted should be brought to a temperature of not less than 55°C and maintained at this for a minimum period of fifteen consecutive days.[14] This is based on the requirements of the US Environmental Protection Agency's Part 503 Rule, which was discussed more fully in the previous chapter. The need to address the potential thermal gradient between inner and outer layers is met within Part 503, which demands at least five complete turnings of the pile during sanitisation. The UK Composting Development Group also stresses the need for the materials to be 'adequately mixed'[15], citing the American Rule as an example, though not explicitly defining a requirement. Specifying these process details is a very valuable approach to the problem, but it is unlikely that consumer confidence, and thus the market, will grow without some kind of standard for the product also. Quality assurance by reference to the production method, rather than an analysis of the final result, is both largely unprecedented and runs counter to common expectation.

A number of regulatory bodies and workers in the field have suggested the adoption of an assessment system based on the prevalence of indicator organisms, rather in the manner routinely found in the water industry. Despite the general European preference for defining parameters on the process, not the product, an EU Experts' Group did advocate such an approach, even proposing the recommended levels of suitable diagnostic microbes shown in table 4.6, as long ago as 1987.[16]

Table 4.6 Suggested Levels for Indicator Organisms in Compost

Indicator Organism	Requirement
Salmonella	Absent in 100g of fresh product
Infective Parasitic Ova	Zero present
Faecal Coliforms	$<5 \times 10^2$/g
Faecal Streptococci	$<5 \times 10^3$/g

While the established process-based approach has merit, perhaps the most useful way forward would be ultimately to consider both production method and product pathogens. The elevated temperature/period of exposure route seems likely to be here to stay, but allying this with some quantifiable measurement of the actual batch pathogen status could produce a truly meaningful standard. In terms of consumer confidence alone, this seems to be a very obvious weapon in the biological waste treatment armoury and, while the requirements may need some further consideration before a final standard can be established, one that will surely see more use in the coming years. There are two principal ways in which this could be brought about, one of which even fits in with the current process requirement bias:

- Set maximum acceptable standards for key indicator organisms in the final product

or

- Establish a required pathogen reduction target to be achieved during composting

As mentioned earlier, the situation for anaerobic digestion, as the other major biological waste treatment technology, is somewhat different, leading some companies to envisage using a final high temperature sanitisation phase for their solid and liquid off takes, either separately or together. Other operators make use of composting as their last step, to mature and stabilise the digestate produced, which, of course, brings the sanitisation of their product back within the discussion of composting in general. However, the potential for the accumulation of pathogens within their process liquor still needs to be taken into account.

In order to maintain and expand the market for biowaste-derived soil additives, some kind of product assurance based on the degree of sanitisation of the actual delivered material, to guarantee its safety and consistency, seems inevitable and it is difficult to understand the resistance to its adoption.

While primary pathogens may affect any of the three groups originally identified, the secondary ones generally tend to represent a risk to the workers at the plant itself, or exceptionally, to those who live or work in proximity to the facility. The most commonly encountered problems of this kind arise as bioaerosols, which principally comprise fungal spores, though bacteria and certain other kinds of microscopic biological particles may also be involved. Direct contact with, or the inhalation of air rich in the spores, can pose potential health hazards, and for obvious reasons this is most likely amongst the workforce. The main likelihood of infection occurs whenever the material is disturbed, especially when the compost is being turned or transported around the site. Once inhaled, these airborne particles, by virtue of their very small size, can penetrate deep into the lungs and the resultant response may vary from mild inflammation, asthma or allergic response to serious infections like aspergillosis.

Though all composts may present these hazards to some extent, the danger from the finished material is far smaller than at the production stage and even here, good management and operational procedures can go a long way towards minimising the risk. Since fungi thrive in conditions marginally dryer than those required for bacterial proliferation, some measure of control can be achieved by careful monitoring of the moisture and humidity levels, particularly for facilities contained within buildings. However, a better and more reliable overall approach is to ensure the use of suitable gloves, face-masks and other appropriate protective clothing, together with proper staff training and an insistence on safe practice, as preventative measures to obviate much of the problem in the first place. The risk posed to the surrounding community by bioaerosols from biowaste composting plants has been the subject of much speculation and a number of studies. These have generally tended to support the view that, provided sensible consideration is given to the local factors likely to affect the potential for a specific site at the planning stage, this issue does not represent a problem. However, the scope for bioaerosol contamination of adjoining sites remains the subject of ongoing research by the Composting Association in the UK and by other bodies elsewhere in Europe and the world.

The persistence within the derived product of potentially hazardous substances, together with other inclusions which may or may not themselves be directly harmful, is the final contamination issue to consider. Broadly they fall into two categories:

- Physical contaminants, like glass, paper, plastic or metal which have become entrained in the biowaste fraction either as a result of incomplete separation or by being accidentally re-introduced.
- Chemical contaminants, typically comprising household or garden substances, like paint, bleach and weed killers.

Leaving aside the additional bulk contributed by physical contaminants, which, dependent on the relative amounts involved, can lead to an increased requirement for processing volume, principally a problem for in-vessel systems, these tend to be either unsightly or potentially hazardous inclusions in the final product. Generally, soil additives produced from biowaste of a source segregated origin tend to contain little paper or plastic, which can detract greatly from the product's appearance when present. For obvious reasons, this affects mixed MSW-derived material far more commonly, as a result of the limited effectiveness of secondary separation. However, this need not necessarily be a problem, since the end use for such products often does not demand that they look perfect. As a landfill cover, or mixed with soil for landscaping purposes in reclamation or restoration projects, for instance, a certain degree of plastic or paper within the mix may be acceptable, particularly if the ground is ultimately to be grassed over. The same argument may also be applied to contamination with stones and metals, and even to some extent, glass fragments, since the likelihood of injury is heavily reduced in large scale professional applications, where literal product handling is rare and most of the material is moved by machine. However, it would be quite unacceptable where there was eventually to be ready public access to the land after restoration and, clearly, in any product intended for the household market. There is some scope for a final screening of the product prior to export off-site, to attempt to remove these unwanted inclusions, but this also represents an additional operational cost. Hence, lower-end product quality tends to be defined by the accessibility and requirements of the intended outlet, though, obviously, further improvement would be possible if an appropriate market premium were to make it commercially viable.

The construction of suitably designed treatment facilities coupled with an appropriate management regime can make a considerable contribution to ensuring minimal physical contamination. Even something as simple as using the same vehicle to collect glass or dry recyclables one day and segregated biowaste the next, without ensuring that the truck is properly cleaned between runs, can take its toll, over time. Sadly, exactly this scenario has been played out, and by more than one cost-conscious authority, seeking to optimise its transport requirements.

The likelihood of chemical contamination arising is highest, again, in biowaste from mixed MSW. However, the incidence of significant amounts of household chemicals entering the waste stream is generally lower than is commonly supposed, since few such containers are thrown away as full or even partially empty, though paint is an exception to this, and moreover, the vast majority enter bins with their lids still attached. Studies have found that, in most cases, those which do spill and contaminate the biowaste are, accordingly, few and far between. In addition, the relatively small proportion of the total waste mass this represents lends itself to a natural dilution and dispersal effect once the mixing necessary for bio-treatment begins.

The potential for garden chemicals to contaminate the biowaste would appear to be much greater, especially under separate collection initiatives. Although some gardeners are moving towards the idea of natural pest control, the majority still rely on a plethora of herbicides, fungicides and insecticides to protect their plants. It would seem reasonable to expect that the likelihood of such chemical residues reaching the biological treatment plant, and thus the final product, would be high, since they are not secondarily introduced contaminants, but have been associated with the plant material itself from the outset. However, the research data suggests that this simply does not happen. For one thing, most of the commonly used examples of all three of these main classes of pesticide have a relatively short lifespan in normal use and laboratory investigations have demonstrated a significantly faster breakdown rate under a composting regime. The in-built delay between initial collection, preparation, processing and the eventual production of the final compost should present any residue with ample time to degrade.

There have been numerous field studies undertaken, particularly in the US, which involved testing for a range of pesticides in source separated 'yard' biowaste, both before and after composting. Although not all have yet published their findings, those for which the results are known have generally shown surprisingly few traces of pesticides in the input biowaste.[17] Of those which were found, the majority belong to the organochlorides, which are a highly persistent group of chemicals and, although many of them have not been in general use for some twenty years or so, traces still exist in the environment, despite the passage of time. Even so, the levels detected were very low, a finding that was also repeated with the smaller number of samples discovered exhibiting traces of currently used compounds. According to these results, source separated biowaste typically arrives at the treatment plant with low to undetectable pesticide concentrations.[18]

The second part of the studies looked at the situation in the composts generated and they found that, with the exception of a few of these organochloride compounds, any pesticide present in the input feedstock either became substantially reduced, or was undetectable in the product. One study into grass cuttings specifically looked at the herbicides 2, 4-D and pendimethalin, having found low concentrations of these in some of its original samples. Neither appeared in the resultant compost[19] and this, together with other investigational data, seems to suggest that herbicides in general reduce from originally low to lower, or even undetectable, concentrations in compost derived from 'yard' biowaste.

There is less direct evidence regarding other forms of biowaste, either from food waste inclusive source separation, or that obtained from mixed MSW. However, the little work which has been done, has found low pesticide concentrations in both and, given the laboratory studies which demonstrate the ready degradability of these garden chemicals under composting conditions, it seems likely that the overall picture will be substantially similar.

It is important not to lose sight of the fact that the potential for contamination by heavy metals, pathogens, pesticides and so on arises only because of what exists in the waste stream in the first place. Biological processing does not in some way increase this, a fact which is sometimes overlooked. Indeed, in absolute terms, separating the biowaste from the rest, ideally at source, or secondarily if needs be, may actually be viewed as a positive contribution to reducing environmental exposure to these materials, particularly in terms of leachate reduction. One thing is certain, the waste, even if not treated to form a compost or other soil product, will still retain its contaminants, only now they are destined for incineration or landfill. We cannot escape the consequences of what we put into our dustbins and to the public eye, can the spectre of emissions to air or accumulation in the ground be any the better option? Clearly, there is a justifiable need for controls on these waste-derived products, but they must be realistic and will probably best be based around the concept of fitness for purpose. Contamination with glass, paper and other inclusions from the waste stream may always make mixed MSW-derived biowaste materials of lesser value, but here too there is scope for a useful contribution to landfill diversion and nutrient recycling, which should not be ignored. Even if, ultimately, mandatory source-segregation becomes the universally accepted norm, as a stage in the evolution of sustainable waste management, such an approach can play a vital role, particularly in those countries with a heavy landfill dependence, which cannot simply make the switch overnight.

It is sometimes said that the furtherance of biowaste treatment is hindered by considerations of scale. Some have taken the view that what is achievable on the laboratory bench cannot be replicated in a commercial setting and, undoubtedly, this is true for some interventions, particularly those calling for large chemical additions, excessive temperatures or mass enzymatic action. Unquestionably such procedures have been shown to work, and they have variously advanced our knowledge, or deepened our understanding of the biological processes involved. They are not, however, commercial options. It is not that they are inherently unrepeatable: it is simply the economics which make them so. Generally, few significant differences would be expected between a small experimental version and full size, provided that the operating conditions were closely replicated. This makes obvious sense, since reactions relying on live biological organisms to bring them about are always essentially functioning at the same scale. Assuming the needs of the microbes are met in the larger system equally as well as they were in the laboratory, then the result should be the same. This is, of course, both the key and the major potential stumbling block; adequately matching these requirements in such a scale-up, while attempting to maintain an awareness of the commercial side of the venture is not often straightforward. Nevertheless, in essence the problem here is one of engineering, not of biology.

Greater difficulty can be encountered when an attempt is made to extend a successful limited pilot scheme, perhaps involving only a few hundred households, to cover a wider base. Once again, the impediments are not biological in origin, but more commonly relate to the logistics of transport, the willingness of the general public and the financial implications of the venture. The issues relevant to moving from a small demonstration scale operation to a full size facility in many ways exemplify those facing the development of the whole of biowaste management. Political will and legislation aside, marketing, disposal and take-up rates remain the limiting factors on biological waste treatment.

The vital importance of effective standards for biowaste-derived products to boost consumer confidence and thereby increase market penetration has already been discussed. As the levels of diversion rise, either locally with the extension of the scope of a particular scheme, or more widely in response to changing requirements and regulations, the overall tonnage of the product made, likewise, increases. This brings with it the need for either an equally improved market outlet, or a proportionately large alternative route for off-site export. In the absence of the former, the latter may mean disposal. Although the idea of diverting material and composting it, simply to send it back to landfill at the end, may seem absurd, while it certainly is wasteful, it does permit the stabilisation of otherwise troublesome biowaste and, furthermore, may offer something of a beneficial use for the product as day-cover. Indeed, this may be the only interim route open to unsightly material derived from less clean sources, like mixed MSW. This is not to suggest that such an arrangement would be viewed as ideal, but as a short-term measure, while the industry establishes itself, it may be a necessity. Just such a role for these types of products has already been mentioned in the context of developing sustainable approaches to biowaste management. To return to the original point, however, any expansion of biological waste treatment, local or otherwise, depends on either genuine end-uses, or disposal, to ensure sufficient outlet capacity to take up what it produces. The proportion of each is itself a factor of the efficiency of marketing and education initiatives, consumer confidence and hard economics.

Those countries which have introduced mandatory schemes for their populations, with persuasive sanctions to guarantee compliance, may well approach full participation. For others, relying on public goodwill and less draconian legislation, take-up rates are frequently less impressive. There are clearly issues of national character and historical background which have a bearing on this, but it cannot be safely assumed that the mere provision of a biological treatment facility will ensure its universal support. The reasons for this are by no means always motivated by malice; a surprising number of otherwise sensible people become hopelessly confused when presented with the typical battery of different containers found at Civic Amenity Sites. In a recent informal study, 46% of the regular recycling facility users asked, had been unsure into which receptacle they should deposit an item of waste, at least once in the previous month. That figure rose to 76% if the preceding six months were taken into account.[20] If nearly half those habitually visiting are experiencing such problems on a monthly basis, and these are people who are informed enough to be making the effort to recycle their refuse voluntarily, the likely confusion of the less waste-aware is easy to imagine. Though there is little published information on this issue for home-based separate collection schemes, it seems unlikely that the same occasional confusion will not occur, though the situation is currently somewhat simplified by generally only having two options – 'dry recyclables' or 'the rest'. This is, of course, of no help to a system intended to derive a clean, segregated biowaste fraction suitable for treatment, though this is not an insurmountable problem, as the experience in Holland and elsewhere clearly demonstrates. However, the difficulty is likely to arise during the transition to such a scheme and it is to some extent summed up by the so-called 'can of beans' question[21]. Should a householder, wishing to throw away an unopened can of beans, put it in the recyclable bin, by virtue of the metal in the can, or the food-waste one, because of its contents? This is clearly not intended to be taken too literally, but it does encapsulate the concern.

Since such considerations are really only of relevance to the collection of biowaste for centralised processing, some local authorities have sought to circumvent the problem by encouraging their population to deal with their own waste, typically by home composting. Providing suitable composter bins, either at no cost to a number of selected households or, more commonly, making them widely available at a subsidised price to any resident wanting one, has many advantages for a local council's overall recycling strategy, which will be examined more fully later. What is germane to the current discussion is that, again, simply providing the means is not to guarantee that it will be used. Indeed, there is some anecdotal evidence to suggest that after an initial flush of enthusiasm, home composting levels fall back to around where they were originally and that within two years, many composter bins lie discarded. Part of the reason for this is the difficulty householders often encounter in obtaining useable compost, which generally results from bad operational practice, which itself typically arises from either poor initial instruction, or an inadequate attempt at carrying out what is required. However, whatever the root cause, it remains clear that high levels of diversion cannot just be taken for granted.

As always, education, marketing and acceptance are intrinsically linked and play a vital role in all aspects of biological waste treatment. A wider awareness of the need for biowaste management is as vital to obtaining a good feedstock as an understanding of the benefits of the product will be to opening up the market. Something in the region of 25 million tonnes of peat is used worldwide for one form of horticulture or another every year, 15 million tonnes of it, in Europe alone.[22] The resultant potential market is huge and the environmental advantages of substituting biowaste-derived materials for this and other kinds of soil additive products are considerable. For one thing, the mineral recycling of the biowaste, via biological treatment, ensures that the ecological cycle is kept as short as possible and furthermore, directly closes the loop between plant waste and new plant growth. Accordingly, the more ancient stores of organic matter, like peat, which are effectively finite resources, can be reserved for those applications for which no suitable substitute exists. Moreover, for those countries which are prone to progressive soil fertility degradation, or erosion, the potential benefits offered by the scale of addition made possible by mass biowaste processing is immense.

Economic Factors

Unsurprisingly, the economics of biological waste treatment are one of the major factors in the whole equation, in terms of both operator and client, and the implications of this to local authority budgeting are very great. Though the specifics may vary under the usual influences of the system used, existing arrangements, taxation regime and locality, the overall costs for biowaste processing are governed by the following four main elements:

- Collection
- Haulage
- Processing
- Product sales

For approaches intending to continue to collect mixed MSW and deliver it to a plant for secondary biowaste sorting, the collection costs will obviously remain the same. Separate collection systems will incur a higher cost, which will represent an increase of between 10 and 65% for normal housing and 20 and 60% for high rise accommodation, based on reported Dutch figures.[23] There is the possibility that these increases can be lessened if the regime can be set up to reduce the frequency of collection, though this has clear implications on the required storage capacity provided. In addition, the public health issues and potential odour nuisance of such an arrangement would also have to be taken into account. There is some evidence to suggest that there are no essential differences in the levels of bacteria and endotoxins in fourteen day-old biowaste, compared with the concentrations present in it at day three, based on analysis carried out on three-, five-, eight- and fourteen-day-old material.[24] It seems likely, however, that public opinion would be against such a switch in areas where routine weekly collections are the norm. Irrespective of how the eventual timings of the collections are arranged, the cost of providing appropriate bins and vehicles, or the modification of existing ones to suit, will be additional initial sources of expenditure during the introduction of source-separation.

It has been suggested that the separate haulage of individually collected biowaste and the remainder of the MSW should not be any more expensive than collecting it in a single mixed batch, since the overall volume remains the same.[25] This may be so in some cases, but it is hard to accept it as universally true. While careful attention to the logistics of transport and the siting of facilities can certainly go some way towards keeping the costs to a reasonable level, the economies of carrying capacity, manpower and time, available to traditional compaction vehicles, are unlikely to be emulated under a separate collection regime. Moreover, from the point of view of an individual authority, perhaps making use of a shared facility with neighbouring councils, the real haulage cost may significantly increase if, for example, the biowaste plant and the landfill or incinerator dealing with the rest of its waste are a distance apart, and possibly in opposing directions.

The cost of the actual biowaste processing is a complex issue. Obviously depending on the technology used, the price to the client authority will represent the running costs of the plant, pro rata on the tonnage supplied, plus the operator's profit margin, unless the facility is municipally run. Even then, it will probably be necessary to generate a return to pay for the initial capital investment, since the authority will have had to build the plant for itself, in the first place. The calculation of the gate fee charged will inevitably reflect the local situation in terms of wage rates, utility costs and so on, but it is also likely to be influenced by the price of alternative waste management options, no matter whether it is run by the local authority itself or by private operators. There will always be the question of what the market can bear and in those countries where landfill has traditionally been cheap, this can form a brake on the uptake of biowaste processing. As landfill costs rise in these lands, as it is anticipated they will in response to increased regulation enforcing stricter standards, the playing field levels and the competition becomes more even. It has been stated that for some countries, composting using relatively uncomplicated systems can be done for around half the cost of landfilling.[26] According to the same report, incineration is between two and four times as expensive as biological waste treatment, if 'environmentally acceptable' conditions are met. Shifting the relative cost-levels by external levy, to force the issue of diversion from traditional disposal routes, is a relatively simple thing for a government to implement and many have begun this process, or are currently contemplating it. In the absence of a market in which biowaste treatment can recover its costs and compete commercially, the only really viable alternative to stimulate its development would have to be direct subsidy, and this generally seems to be a less popular form of intervention.

The question of product sales is bound up with the issues of standards, consumer confidence and available markets, which have already been discussed at some length. Obviously, when a revenue stream is generated from the sale of compost or energy (in the case of anaerobic digestion) the net cost of biowaste treatment falls, which, in turn, helps to increase the venture's commercial viability. However, a scheme which relies entirely on the sale of products for its survival must be regarded with some suspicion, if only because over the likely lifetime of the plant, it will be completely vulnerable to the vagaries of consumer demand. Sales are welcome additions to the cash-flow; they are best not viewed as financial guarantees. As experience has repeatedly shown with other recycled products, perhaps most notably paper, the market is habitually fickle. There is no doubt that revenue from good quality biowaste products can be a very valuable contribution, but there is an associated cost in obtaining the high standards. As with all of biological waste treatment, it is a question of finding the optimum combination.

As long ago as September 1994, the Organic Reclamation and Composting Association (ORCA) put forward a set of guidelines for biowaste processing in Europe, which were specifically devised to be pertinent to all technologies and appropriate applications. ORCA's sister organisations, the relevant Composting Councils in America, Canada and Japan are now involved in examining how these same principles can be extended to their own respective countries and local conditions. The move towards establishing widely-accepted standards in bio-treatment practice has begun, and it is firmly rooted in the twin concepts of sustainability and integrated waste management, emphasising the essential imperative of achieving equilibrium between diversion from landfill, operational cost and product quality. These are the three fundamental and competing elements of effective biological waste management.

In the final analysis, biowaste treatment must meet four simple criteria and the success or failure of an individual plant, technology or operator may be defined by how proficiently this is accomplished.

1. The process must be capable of bringing about the full and proper treatment of all the biowaste it accepts.
2. Environmental Safety must be ensured throughout the collection, processing and final use or disposal of the product(s).
3. The final product should be of sufficient quality and suitably free of contamination to be acceptable for its intended purpose.
4. The process should increase the overall diversion of biowaste from landfill.

Having taken an overview of biowaste treatment in this section, the following chapters will examine some of the specific ways in which these four goals can be achieved.

References

1. DHV Environment and Infrastructure BV (Amersfoort), in cooperation with Plancenter Ltd, (Helsinki) and University for Soil Management, (Vienna) *Composting in the European Union*, a final report to the European Commission DGXI, Environment, Nuclear Safety and Civil Protection, 1997.
2. Ibid
3. *A Way with Waste; A Draft Strategy for England and Wales*. Part Two, Department of the Environment, Transport and the Regions, 1999, p. 90.
4. Diggleman, C. and Ham, R. K., Life Cycle Comparison of Five Engineered Systems For Managing Food Waste, Final report to the National Association of Plumbing, Heating, Cooling Contractors, University of Wisconsin, 1998.
5. *A Way with Waste; A Draft Strategy for England and Wales*. Part Two, Department of the Environment, Transport and the Regions, 1999, p. 90.

6. Author's own unpublished data.
7. Landell Mills Market Research, cited in *Marketing Guide for Producers of Waste Derived Compost*, ADAS Woodthorne for the Department of the Environment, Transport and the Regions, 1997, section 3.9, p. 24.
8. DHV Environment and Infrastructure BV (Amersfoort), in cooperation with Plancenter Ltd, (Helsinki) and University for Soil Management, (Vienna) *Composting in the European Union*, a final report to the European Commission DGXI, Environment, Nuclear Safety and Civil Protection, 1997.
9. Gruneklee, C. E., *Development of Composting in Germany*, the Proceedings of Orbit 97, Zeebra Publishing UK, 1998, cited in Stentiford, E. *The Supermarkets are Coming*, Wastes Management, The Monthly Journal of the Institute of Wastes Management, January, 1999, p. 24.
10. *Product Guide for Compost Specifiers*, ADAS Woodthorne for the Department of the Environment, Transport and the Regions, 1997, section 3.4.4, p. 12.
11. *Anaerobic Digestion*, Institute of Wastes management, 1998, p. 33.
12. Ibid.
13. Border, D., *Composting in the UK and the Rest of Europe*, Wastes Management, The Journal of the Institute of Wastes Management, August 1997.
14. *Report of the Composting Development Group on the Development and Expansion of Markets for Compost*, presented to Ministers for the Environment, Department of the Environment, Transport and the Regions secretariat to the Composting Development Group, July 1998.
15. Ibid
16. Zucconi, F. and Bertoldi, M., *Compost Specifications for the Production and Characterisation of Compost from Municipal Solid Wastes*, in Compost: Production, Quality and Use, Elsevier, 1987, pp. 30–50, cited in Stentiford, E. *The Supermarkets are Coming*, Wastes Management, The Monthly Journal of the Institute of Wastes Management, January, 1999, p. 25.
17. BioCycle Magazine, September 1998, p. 16
18. Ibid
19. Ibid
20. Author's own unpublished data.
21. Pell, I. E., personal comment.
22. DHV Environment and Infrastructure BV (Amersfoort), in cooperation with Plancenter Ltd, (Helsinki) and University for Soil Management, (Vienna) *Composting in the European Union*, a final report to the European Commission DGXI, Environment, Nuclear Safety and Civil Protection, 1997, p. 35.
23. Ibid, p. 30. (Section 5.2.2, s.v. <u>Economic implications</u>) 'For low-rise buildings these costs may increase from 45 ECU per ton for collecting the waste integrally to between 50 and 75 ECU per ton for separate collection. . . . For high-rise buildings the increase will be from 25 ECU to between 30 and 40 ECU.'
24. Andresen, S. and Rasmussen, K., *Source Separation and Collection Systems for Biological Waste Collected from Households*, Proceedings of Biowaste '92, ISWA/DAKOFA publication, 1992.
25. DHV Environment and Infrastructure BV (Amersfoort), in cooperation with Plancenter Ltd, (Helsinki) and University for Soil Management, (Vienna) *Composting in the European Union*, a final report to the European Commission DGXI, Environment, Nuclear Safety and Civil Protection, 1997, p. 30.
26. Ibid, p. 31.

CHAPTER 5
Composting

There has been some confusion in biological treatment terminology in recent years, with 'composting' being applied in certain circles to all forms of biowaste breakdown, anaerobic as well as aerobic. Thus, when these workers have spoken in this way, it has led to more than a few misunderstandings. Throughout this discussion, the term 'composting' will refer solely to the exothermic decomposition of biodegradable materials, in the presence of oxygen.

The natural breakdown of such matter is a well appreciated phenomenon, widely used by gardeners, horticulturists and the like for centuries to provide a relatively stable, nutrient-rich compost of a kind suitable for use in plant growth or on the land, for soil improvement. The burgeoning changes in legislation and waste management practice, driven, as discussed previously, by the need to reduce the traditional problems associated with biowaste under traditional landfill arrangements have combined to make the treatment of this biodegradable fraction of refuse increasingly important. Accordingly, composting, particularly at a large scale, is receiving increasing interest as a means of biowaste treatment. Dealing with the large quantities of biowaste typically derived from household involves a much larger scale of operation than the traditional back garden compost heap and this has certain implications for these operations, particularly in materials handling, residence time and the oxygenation of the decomposing matrix.

The principle of providing large-scale controlled conditions to bring about the managed aerobic decomposition of biowaste, and the production of a safe and stable end-product is well established and a number of commercial composting systems exist which encourage aerobic decomposition in various ways. However, before looking at these systems in any great depth, it is worth first considering the composting process itself.

Organic matter generally breaks down more efficiently and completely in conditions of ready oxygen availability, largely as a result of the plentiful free energy resulting from aerobic respiration, which uses the oxygen as the final electron-acceptor. Under these circumstances, in overview, proteins are degraded to nitrogen or ammonia and ultimately mineralised to nitrate, while fats and carbohydrates break down to carbon dioxide and water, via organic acids. Though this represents the macroscopic mass flow, at the microscopic level, of course, a proportion of the material degraded becomes incorporated into microbial cells, as the decomposer microorganisms themselves grow and multiply. Even under idealised conditions, there are a number of rate limiting steps in the composting process, including the release of extracellular hydrolytic enzymes, the speed of hydrolysis itself and the ease of oxygen transfer, which depends on physical factors related, amongst other things, to the particle size and nature of the biowaste material.

The specific kinds of biowaste suitable for treatment by aerobic composting can vary greatly, particularly that which arises from municipal solid waste. As has previously been mentioned, seasonal variation, local conditions and climate all may play their part in providing a highly heterogeneous feed material. On the other hand, for certain applications of composting, for example, food processing or horticultural biowastes, the materials to be processed may approach homogeneity. Accordingly, the precise biochemistry of breakdown may be very complex, involving many intermediary compounds and progressing via many agents along a number of biological pathways.

However, aerobic composting can be typically characterised as proceeding through the following four distinct general phases, which are chiefly determined by their temperature characteristics, as shown in the following figure.

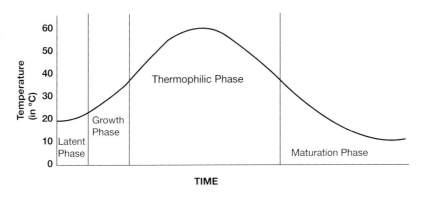

Figure 5.1. Temperature and the phases of composting

The Four Phases of Composting

i) Latent phase (Ambient temperature – c.22°C)
Acclimatisation, infiltration and colonisation of the material by the bacteria, fungi, protozoa and other micro-organisms responsible for composting.

ii) Growth phase (c.22°C – c.40°C)
Micro-organisms grow and reproduce, leading to high respiration rate and elevation of temperature to mesophilic range.

iii) Thermophilic phase (c.40°C – c.60°C)
Peak temperature is achieved during this phase, with maximum pathogen sterilisation taking place. Temperature drops to around 40°C at the end of this phase.

iv) Maturation phase (c.40°C – ambient)

A slow, secondary mesophilic phase, subsequently dropping to ambient temperature as biological activity within the material decreases. Complex organic chemicals are transformed into humic compounds and residual ammonia undergoes nitrification to nitrite and subsequently nitrate.

The scope for artificial environmental manipulation within the composter to enhance biological breakdown in the desired manner lies chiefly in the reduction of the duration of the first, or latent, phase. Commercially, the more rapid the acclimatisation, infiltration and colonisation of the material, the sooner the composter will be available to take a renewed load of biowaste.

As has been discussed in earlier chapters, one of the principal objectives of biowaste treatment is the stabilisation of the material, thereby rendering the originally putrescible organic waste into largely inorganic forms, which are relatively inactive and thus pose no significant risk of polluting the environment. The composting process, while bringing this about, also causes a linked reduction in the carbon to nitrogen (C:N) ratio as quantities of organic carbon are converted to carbon dioxide. This is an important consideration, since a C:N ratio much above 25:1 can inhibit nitrogen mineralisation and proper final maturation of the compost, which has obvious consequences for any intended use of the end product as a fertiliser or soil enhancer. Accordingly, where mixed-source waste is taken for composting, it may often be necessary to ensure a good degree of feedstock mixing and blending to produce a C:N ratio appropriate for effective processing. The following table gives illustrative C:N values for some typical biowaste forms and demonstrates that it is also possible to characterise biowastes as either 'High Nitrogen' or 'High Carbon' materials.

Table 5.1 Illustrative Typical Carbon to Nitrogen Ratios

Material	C:N	
Wood and Sawdust	500:1	
Paper	170:1	HIGH
Bark	120:1	CARBON
Leaves and Foliage	60:1	MATERIALS
Horse Manure	25:1	
Cow Manure	20:1	HIGH
Grass Clippings	19:1	NITROGEN
Sewage Sludge (digested)	16:1	MATERIALS
Food Wastes	15:1	

The bacteria responsible for nitrification belong to two genera; *Nitrosomonas*, which converts ammonia to nitrite and *Nitrobacter* which completes the mineralisation by turning this nitrite into nitrate. Since nitrate is the form in which most plants take up their required nitrogen, a suitable original feedstock C:N ratio coupled with an adequate maturation phase are essential prerequisites in the production of beneficially usable compost. Their activity is effectively confined to the maturation phase since they have a relatively slow growth rate and are inactivated by temperatures in excess of 40°C. Hence, only after the growth and thermophilic stages of the composting process is the way clear for these organisms to play their part.

Compost Biology

The organisms involved in the composting process fall into three main groups:

- Micro-flora (actinomycetes and bacteria)
- Meso-fauna (mites, nematodes and protozoa)
- Macro-fauna (beetles, earthworms and millipedes)

Some authorities classify these organisms either into primary, secondary and tertiary feeders or in terms of how they obtain their energy and carbon. Hence, all of the examples cited previously in brackets may be termed heterotrophic, since they get their energy and carbon from the direct ingestion and oxidation of organic matter. Autotrophic organisms, like nitrifying or sulphur-oxidising bacteria, which obtain their energy from non-living sources, as in photosynthesis, and their carbon directly from atmospheric carbon dioxide, also play an important role in the composting process.

During the four phases of the composting process, there is an allied biological succession of organisms acting on the biowaste being treated. In the initial latent stage, a variety of bacteria, fungi and other microbes become acclimatised to the particular waste material to be composted, gradually infiltrating and colonising as the temperature rises from ambient. As their activity increases and the temperature continues to rise as a result, mesophilic organisms, and bacteria principally amongst these, are the first to begin the decomposition, until the increasing heat shifts conditions in favour of thermophilic forms, which proliferate through the composting matrix and play a significant part in the breakdown of the proteins and carbohydrates present. Above 70–75°C, however, even these heat-loving organisms become inhibited. Later, the falling temperature permits actinomycetes to become the predominant biological group throughout the final stages of composting, accounting for the characteristic whitish-grey surface colour often seen on ageing compost heaps. Although mainly confined to the outer edges of the pile, they play a vital role in the process, which is particularly important in the context of the centralised treatment of MSW and garden waste, since they decompose the cellulose and lignin, which are the less readily biodegradable components of the biowaste.

The microbiological component of compost has received increasing attention over recent years, particularly as a means of evaluating the likely worth of the finished product to users. Traditionally, compost analysis has tended to concentrate on the standard NPK values, treating the material in much the same way as a fertilizer. However, it has become clear that compost is a much more complex substance than fertilizers and can offer potential benefits to a grower or gardener that far outweigh its mere mineral contribution to a soil, particularly as a suppressant of certain plant diseases and as an inoculant for microbe impoverished ground. Accordingly, in some areas, both producers and users of compost are beginning to use the idea of a microbiological profile as a means of compost characterisation and assessment. The extension of soil microbiology assay techniques to composted materials has been pioneered in the US by the BBC Laboratories in Tempe, Arizona, where a standard analysis is performed based on the concentration of what they have termed six 'functional groups' of micro-organisms. Thus the compliment of aerobic, anaerobic and nitrogen-fixing bacteria, fungi, actinomycetes and pseudomonads present in a sample forms a predictive tool for the value of the material as a soil microbial inoculant[1]. As mentioned in the previous chapter, poor quality soils can be improved dramatically by the judicious addition of a compost with a thriving community of micro-organisms.

According to this work, adequately matured compost will have between 10^8 and 10^{10} Colony Forming Units (CFU) of heterotrophic aerobic bacteria present per gram of dry weight. Below this figure, the material's potential soil inoculant function is reduced and there is evidence to suggest that it may also significantly impair its ability to suppress plant diseases. The aerobe: anaerobe ratio should be no less than 10:1, a lower figure indicating a too-infrequent turning regime at the production facility and would highlight the need for further treatment of the material prior to use for plant growth or germination, to remove possible residual anaerobic by-products. Actinomycetes, it is suggested, should be present in concentrations of between 10^6 and 10^8 CFU/g (dry weight). Since these organisms have many functions, especially in breaking down complex structural materials, they may be evident in higher levels in compost derived from the woodier kinds of biowaste. Being particularly effective in alkaline soils, their ability to improve crumb structure and their strong involvement in disease suppression makes their presence in properly matured compost essential. The population of nitrogen-fixing bacteria varies considerably, being linked to the available nitrogen present. There is an inverse relationship between the amount of biological nitrogen in the compost and the numbers of free-living nitrogen-fixers found, since their concentration increases as the available nitrogen decreases. A range of 103 and 106 CFU/g (dry weight) is hypothesised as typical. Pseudomonads, which play a vital part in nutrient recycling, particularly in the regulation of phosphorus availability, should generally appear in the same range as the previous group, though in certain circumstances, which are largely dependent on the original feedstock, there may be fewer. They have also been implicated in the biological control of plant disease, making their presence of some significance in the finished product.

The final group, fungi, though represented in relatively smaller numbers at around 10^3 to 10^4 CFU/g (dry weight), are of considerable importance since, in addition to their role in biodegradation, they also perform a wide variety of tasks *in situ*, including disease suppression, soil stabilisation and nutrient recycling.

It has been suggested that many of the problems experienced by users with the application of biowaste-derived composts arose as a result of the material itself being immature. Compost maturity is largely a measure of the absence of certain phyto-toxic compounds which are typically to be found in biowaste undergoing treatment. Hence, their presence in a final compost is characteristic of an unfinished or poor quality product and may lead to the inhibition of germination, rapid nitrogen depletion, root damage and even plant death in extreme cases. As has been discussed previously, the lack of an agreed standard for this and other areas of compost production helps neither consumers nor producers. However, compost quality issues such as resident microbial community, maturity, stabilisation and mineral content, and, perhaps most crucially, the interplay between them, are becoming more widely appreciated. Moreover, the use of parameters other than the traditional fertiliser-based ones to obtain a more complete compost profile, represents a major advance towards the potential wider applications of biowaste-derived materials. It is unrealistic to expect that all such materials, produced from the widely differing feedstocks available and by the many different systems and management regimes commercially employed, will be able to compete for a single market niche. There is, as has been mentioned previously, significant scope for considering a *fitness-for-required-purpose* approach, rather than viewing all biowaste-derived material against a single universal standard. Characterising these composts more completely in terms of their microbiology, stability, maturity and similar factors, in conjunction with the usual nutrient content analysis, provides the first real step in matching the resource as produced, to intended end use.

Composting as a Method of Biowaste Treatment

Composting itself is an *extensive* technology, as defined by the Dutch Research Group NTO, being a simple, low intervention and slow-acting approach, with a fairly low demand for resources and relatively modest initiation, running and support costs. It has, accordingly, a clear appeal to local authorities, charged as they are with simultaneously honouring their obligations regarding biowaste and balancing their budgets. Unsurprisingly, many of the initiatives which have been set in place, or are proposed, to deal with the growing requirements of legislation governing the disposal of biowaste have encompassed composting as a central facet. While there are many factors which can tend to influence the specific details of schemes in one way or another, in general terms, the application of composting to the biowaste element of MSW and allied wastes falls naturally into one of two options, namely, home composting and centralised facilities.

Home Composting

Home composting invariably calls for the direct involvement of the householders, who are required to remove the biowaste component from their general MSW and consign it to specially constructed compost bins. These bins are sometimes provided at no cost by the local authority, typically as part of a small trial scheme, or at a subsidised price where the aim is to increase the use of home composting over an entire authority's area, without formalising it. A number of UK authorities have held such sale days and huge quantities of composters have been purchased. On the face of it, at least, the idea of a do-it-yourself approach seems popular, both with the councils and with the public.

However, the matter is not without its concerns. One of the most commonly encountered problems with home composting is the selection of the bin itself. There is quite a variety of designs available and many undoubtedly work well. It is as unfortunate as it is inevitable that the cost-conscious consumer – be that householder or council – will tend to be drawn to the budget-end of the market, where certain of the composters require considerable effort to be expended by the user if they are to work properly, if at all. The provision of poor quality bins, coupled with an unwillingness or inability to meet the extra demands of running them, is an often repeated pattern in the downward spiral of compost abandonment. By contrast, those individuals benefitting from well-designed composters, who produce good quality material at the end, are far more likely to persevere with necessary waste segregation required to make the scheme work. It is not a difficult concept, but it does seem to be one ignored more frequently than might reasonably be expected.

The uptake of home composting also depends on the ability of householders to make use of the product they make, which acts as an obvious bar to the universal application of this approach to biowaste treatment in areas which have their share of high-rise or similar properties. Clearly, there is little incentive for anyone without a garden to turn out compost. While the principal advantages of home composting are its low initial cost, speed of implementation, 'feel-good' public involvement and instant potential increase in landfill diversion, certain experiences of actual scheme management would seem to suggest that after the initial enthusiasm wanes, in many cases active home composting levels tend to tail off to their original or near-original levels. Anecdotal evidence from a variety of sources[2] indicates that, at least in some areas, many of the bins bought or supplied lie unused within eighteen months to two years of acquisition. Undoubtedly, much of the reason for this relates to the ease-of-use of the bin provided, as has been mentioned previously, but it also amply demonstrates the heavy dependence of home composting as a biowaste treatment strategy on householder goodwill and competence. Simply making the means available is not to guarantee its use and, hence, many of the diversion figures claimed for areas operating such schemes are viewed with some degree of suspicion, simply because they are effectively unverifiable. For this reason, in the UK, the Audit Commission will not accept amounts of waste postulated as having been diverted from landfill via home composting schemes towards the required performance indicators on recycling for Citizens' Charter purposes. There is a general scarcity of published studies which have generated actual objective data on this issue, but certainly some of what has been made available would indicate that the general perception of home composting as a means of waste minimisation/diversion is not an accurate reflection of the true situation.

One such investigation looked at a trial scheme in the Borough of Luton[3] and was somewhat unusual in that the amount of waste requiring disposal in the traditional manner was weighed over an eight-month period for both the trial area and an additional control area for comparison. These samples consisted of the same number of houses, of similar size and type and located in similar socio-economic neighbourhoods. From the outset, the results were surprising, with only 40% of households in the trial area (86 out of 217) taking up the offer of a free composting bin as opposed to the 60%+ initially predicted[4]. Moreover, in a follow up interview campaign, the actual number of properties actively participating was found to have fallen to only 35%. At the end of the eight-month period, when compared with pre-scheme baseline figures, the results for the trial area indicated an 11% increase in waste arising, while the control area had reduced its output by 8%[5]. In absolute terms, these variances may not be especially significant, as a relatively small sample looked at in the context of the whole Borough's waste stream. However, what they would seem to suggest is that home composting, *per se*, may make little actual difference to the overall amount of waste generated and it certainly does not automatically guarantee the kind of instant minimisation popularly supposed.

A similar study in New York City, which targeted four residential neighbourhoods, one each in Brooklyn, the Bronx, Queens and Staten Island (Manhattan itself having been left out due to the preponderance of high-rise apartment buildings) reached a broadly similar final conclusion. However, while backyard composting was thought unlikely to make a significant contribution to landfill diversion, the city's Department of Sanitation reported heightened awareness of waste issues and recycling in their pilot scheme areas[6]. As a result of their own experience[7], Luton Borough Council decided not to implement a wider policy of home composting, choosing instead to establish a central collection system at their household waste recycling centre, which is reported to be working very successfully and diverting a 'good percentage of the waste entering the site'[8] away from landfill and into composting.

Another area of concern, though one less widely appreciated, for home schemes is the potential for pesticide persistence in such operations. The ability of centralised composting treatments largely to eradicate herbicide and similar garden chemical contaminants was described in the previous chapter. While there has not been anywhere near the same level of investigation into this aspect of domestic composting, there has been concern raised in some quarters regarding the possible accumulation of pesticide residues over time in small scale, effectively closed systems. For one thing, herbicide-treated plant material entering a home composter bin is not mixed with additional biowaste from other sources, as in a large-scale operation, which obviously denies it the feedstock dilution which occurs naturally in a municipal facility. Moreover, since a number of studies have implicated grass clippings as the major herbicide-carriers in the domestic garden waste stream, the large preponderance of this type of material entering home composting over the year has clear implications for at least the possibility of compost contamination. In an attempt to derive some kind of indication of the likely levels which would be detrimental in use, one investigation deliberately introduced herbicides into finished compost, adding MCPP, dicamba and 2, 4-D to material used for growing tomatoes. Based on these findings, it was recommended that the final concentration in compost should be less than 0.1% of the level normally found in grass clippings one day after treatment with these herbicides, at the typical manufacturers' recommended rate of application. How fair a reflection this is of the true post-composting situation remains to be determined, but it does suggest that the issue is one which should not be readily discounted.

However, householders do have one clear advantage over the operators of centralised sites in that they have a much more intimate knowledge of what they have actually put into their bins and are obviously far better able to dictate the amount and type of chemicals used on their garden and, thus, likely to enter their compost. This degree of control is, of course, impossible at a municipal facility and may go some way towards balancing out the inbuilt dilution advantage mentioned in respect of these sites. This again serves to highlight the inescapable dependence of home composting on the participant, requiring willingness, diligence and a certain level of householder awareness to succeed. Clearly, there is a place for home composting as a part of a local authority's response to the need for biowaste treatment, but it seems unlikely that it could ever make the sort of significant inroads into the problem which will be required by the EU Landfill Directive, or other comparable legislation around the world, to be any authority's sole, stand-alone approach. Where such schemes can make the best contribution may well turn out to be in those areas which for reasons of remoteness, either in absolute terms or simply relative to the location of the centralised plant, are less readily serviceable by the kind of separate collection regime likely to be required for a typical municipal operation. This has the additional advantage that, particularly in rural or semi-rural areas, there is often a well-established existing background acceptance of composting and it is not difficult to integrate a more formalised strategy into such a cultural context.

Centralised Composting

Collecting biowaste and composting it at a centralised site removes many of the problems of the previous approach, though not without introducing some new ones. The provision of facilities large enough to meet the needs of effective waste management on the municipal scale generates its own specific operational considerations. Though the underlying principles of the biochemistry and microbiology of all composting processes are relatively uniform, irrespective of application size, the physical issues of dealing with the kind of volumes involved, impose their own restraints.

The importance of adequate oxygenation of the composting material is well known, since the proper aerobic breakdown of biowastes can only take place if the micro-organisms responsible are provided with a sufficient supply. Areas within the material denied oxygen become anaerobic, decomposition ceases to be exothermic and slows down or ceases altogether, with bad odours characteristically arising as a result. This tends to be avoided in a traditional gardener's compost heap, or in the home composter, because of the relative ease with which oxygen can diffuse directly into the degrading waste. However, in a commercial operation, the larger amounts of waste involved result in a lower surface area to volume ratio, which becomes the limit to natural diffusion to the central core. In order to overcome this problem, centralised composting relies either on some system of mechanical turning of the decomposing material to aerate it, or the direct pumping of air through the matrix. The merits of these rival approaches will be examined later, but whichever is employed, the external energy requirement to effect adequate oxygenation has its own cost implications for the commercial side of the operation as a viable biowaste management solution.

It is well known that garden waste is particularly prone to seasonal peaking and, consequently, a dilemma arises as to whether the composting plant should be designed to take the peak flow, knowing that for half the year it will be under-utilised, or sized to the average rate arising. The latter, of course, means that during the summer months some stockpiling of biowaste is inevitable, which, though possible, is generally taken to be undesirable, commonly being viewed as a potential health hazard, a source of bad odours and a guaranteed way of attracting vermin. In the context of a large-scale operation under the aegis of a local council or other, similar authority, such a proposal might well be found to be politically unacceptable, unless the plant is situated some distance from population centres.

In general terms, municipal facilities require a certain amount of ancillary equipment such as shredders and specialist installations or devices as dictated by the particular method being used and a workforce element, together with an area of suitable land and buildings. The actual physical extent of the operation will obviously depend on the amount of waste to be composted, the technology to be used and any desired site combination with wider recycling or waste management initiatives. Additionally, consideration will have to be given to the final destination of the finished product, since the scale of such municipal operations inevitably means a sizeable and regularly arising quantity of compost will be routinely generated. Obtaining a suitable guaranteed outlet, by effective marketing or otherwise is, therefore, an essential prerequisite for any such scheme.

The feedstocks for centralised operations are typically relatively pure, customarily arriving from the garden/yard waste fraction deposited at Civic Amenity Sites, via separate collections of segregated 'green' or 'food' waste, direct from the householder or similar kinds of sources, as illustrated by the following table.

Table 5.2 Sources of Green Waste for Composting in the UK, 1997

Source	Tonnage Composted
Local Authority Parks and Gardens	15,800
Kerbside Collection of Household Garden Waste	28,280
Non-Differentiated Landscape Park and Household Green Waste	73,000
Household Garden Waste from Civic Amenity Sites	343,964

Source: A Way with Waste[9]

Thus, for reasons previously discussed, the material is ideally suited for treatment by biological means.

There are sites across the world which accept mixed MSW, relying on some form of manual or mechanical separation regime to produce a biowaste-rich fraction for composting, but this approach has not found widespread favour and does not represent current mainstream thinking. For those operations which do follow this route, generally the idea has been to separate the biowaste by default, using a variety of established techniques to remove the other constituents of the waste, working on the principle that what then remains is largely the desired material. Dependent on the system employed, the composting technique to be used and the eventual end market of the compost, there is some merit to the approach, especially where a high quality product is not required. Whether such methods can be applied universally remains a matter of some doubt, since there can only be a limited outlet for these low grade materials and though some have claimed remarkable successes for integrated mechanical MRFs, as has been discussed before, the evidence of true operational viability has often been conspicuously lacking. It seems likely, then, that there is little real advantage to be gained in attempting to sort mixed MSW by any more complex means than those already available and widely known. The simple approach of separation by size, typically taking the 'fines' from mechanisms no more sophisticated than a standard trommel or screen, would empirically appear to yield a waste fraction good enough for what might be termed 'dirty' composting. A fully mechanical MRF to do more than this seems likely to remain elusive, and to some extent, unwarranted, at least for the foreseeable future.

Commercial-scale composting systems in general use fall into five main categories:

- Windrow
- Static Pile
- Tunnel
- Rotary Drum
- In-vessel

There is a sixth type, namely tower composting, which is sometimes encountered, but far less commonly than the other forms mentioned above. The most appropriate system for any given application will largely depend on land availability, the nature and quantity of biowaste to be treated, financial considerations and available workforce, the quality of the end product required and the amount of time available for processing.

Windrow

Windrow composting requires the biowaste material to be formed into a series of parallel long rows, typically trapezoidal in shape, often two or three metres high and three or four metres across at the base. Although they can be successfully contained within a building, the scale of the schemes generally using this technique usually dictates that they are open-air facilities, which consequently makes precise process control difficult, since they are variously prone to drying out or becoming saturated, dependent on the weather. However, for composting the kinds of biowaste commonly found in household, park and garden refuse, this need not present much of a problem, as this material biodegrades reasonably readily within a fairly broad range of treatment parameters. Some of the early applications were dogged by excessive leaching in conditions of heavy rainfall, which had been known to cause localised soil pollution, but the now near-universal standard requirement for a concrete pad and a liquor interceptor has largely made this a thing of the past.

Although some aeration is achieved by the natural action of convection currents and diffusion, the windrows are mainly aerated by regularly turning, usually by specialised machines, which also helps to mix and homogenise the composting matrix, thereby further optimising conditions for accelerated decomposition. Front end loaders may be used to perform this function at smaller sites, which cannot justify the expense of the dedicated turners, or at larger ones as an additional measure, but they cannot mix the material as thoroughly or efficiently, which inevitably affects the overall process performance. The turning machines are designed to straddle the body of the windrow and move along it, the larger ones being self-propelled while smaller units are driven by a PTO and pulled behind a tractor. A variety of designs are in common use, achieving the desired effect by means of tines, paddles or similar devices attached to a rotating drum or elevating face which lift, aerate and mix the material. One obvious consideration which must be borne in mind is that the dimensions of the windrows constructed must match the design capacity of the chosen turner. During the initial period of high oxygen demand, windrows are turned frequently, the intervals generally becoming longer as composting proceeds, though dependent on the type of biowaste being processed, turning may be required up to three or more times per week.

Windrows account for the vast majority of centralised composting operations, despite their drawbacks, principally in terms of their very high land requirement, potential for odour problems and concerns over the release of bioaerosols during the regular periods of turning. However, since this form of composting is often carried out as an adjunct to an existing landfill operation, the actual nuisance is frequently significantly reduced.

Static Pile

In broad terms, these may superficially resemble windrows in appearance, but since, as the name suggests, they are not turned, or only infrequently, they are free of the constraint of having to conform to the dimensions of a turner. Thus, the rows can often be considerably taller and wider than in the preceding system. What turning is required is usually done using a front-end loader, which is adequate to meet these needs on the few occasions a year it is necessary. As an obvious consequence, static pile systems tend to be cheaper in respect of equipment, manpower and running costs, but inevitably demand more land because the decomposition progresses at a slower rate, leading to the material remaining on site for a longer period.

A variant on the static pile idea makes use of forced-aeration, and has been successfully used in certain applications, most notably for the co-composting of food or garden biowaste with sewage sludge or animal manures. This requires the composting to take place on a perforated floor, or on top of a series of perforated pipes running at the base of the pile, which are connected to a fan or pump which forces air through the biowaste under positive pressure, or draws it through under negative pressure. While it overcomes the characteristically low oxygen levels within the core of traditional static piles, which accounts for their slower processing time, the bulk movement of air in this way is an expensive method over time. However, for some kinds of materials or for relatively modest facilities accepting only a small tonnage of biowaste for processing, particularly where the need for effective odour control is of major importance, the aerated pile approach has considerable merit, since at this sort of scale, it can be housed relatively easily within a building.

Tunnel Composting

Adaptable for either under-cover or outdoor installation, this system has been used in the mushroom industry for many years and has also begun to receive increasing attention for composting MSW-derived biowaste. The processing takes place within closed tunnels of varying construction, typically forty metres or so in length and four or five metres high. Some of the recent household waste applications have employed a proprietary system akin to a series of one metre wide, sixty-metre long polythene sausages, into each of which a purpose built hydraulic filling machine packs around seventy-five tonnes of source separated compostable material. This design of system is something of a hybrid since it also utilises a fan driven aeration pipe to oxygenate the material rather in the manner of the previous technique, while strategically placed slits in the side wall permit the excess carbon dioxide to escape.

The environmental conditions within tunnels can be more readily controlled than in the case of an aerated pile reactor, which makes the processing time shorter by comparison, though the material usually requires an additional period of post-composting maturation before it can be exported for use.

Rotary Drum

This kind of composting system has been widely used throughout the world at various times but seems destined to follow a perpetual cycle of falling into, and then out of, fashion. The biodegradable material is placed inside the body of the drum, which then slowly rotates. This action encourages the decomposing biowaste to tumble gently and so helps both to aerate and mix it, with some designs having internal paddles or fins fitted to increase this effect. The composter drums themselves are usually steel fabrications and are generally insulated to reduce heat loss. Like aerated piles, drum systems have found some favour with operators seeking an effective means to achieve the co-composting of mixed types of biowaste, most particularly sewage sludge with the addition of a more fibrous material, such as crop residues, straws and garden waste.

In-vessel

Sometimes known by the alternative title of Closed Reactor Composters, in-vessel systems process the waste in a partially or wholly enclosed container, thereby permitting the internal environmental conditions to be closely regulated. There are a number of designs in use, ranging from small steel or plastic tanks, through larger mesh cages to long concrete constructions with high sidewalls to retain the compost and support the ancillary equipment required. The operating principles are essentially the same as for windrows or aerated piles in that the body of biowaste being treated is aerated either by being mechanically turned, as in the case of the larger concrete channels, or by having air forced through it, but the individual units themselves are much smaller. While this approach makes very efficient use of space and offers a considerable degree of precise process control, it is significantly more expensive on a tonne for tonne basis, which obviously makes it less suitable for large capacity requirements. However, such systems' inherent fine parameter control lends them a high degree of flexibility with respect to input waste types, which effectively guarantees their usefulness in certain applications, especially where the material requiring treatment arises in relatively small quantities and does not easily fit into other kinds of process arrangements.

In-vessel systems employing a mechanical turning regime normally use self-propelled specialist machines which aerate as they travel along rails fixed to the side walls. Typically equipped with either a rotating belt fitted with lifting tines or a revolving shaft mechanism with blades at the head, they are designed to move the composting material down the length of the vessel until it is ultimately expelled as the finished product. Dependent on the nature of the material itself and the specific design and operation of the vessel, this may take as little as a few weeks, though, again, some secondary maturation may be required. These facilities are customarily sited within a building or at least under some form of roof structure, since they need protection from the rigours of the elements. Being an essentially artificial class, defined by the 'where' rather than the 'how' of processing, there is a marked variance in both the capacity and complexity of available systems and, accordingly, also in their cost.

Although the site specifics will vary for obvious reasons, the operational layout of centralised composting facilities share a number of generalised common features, as shown in the following diagram.

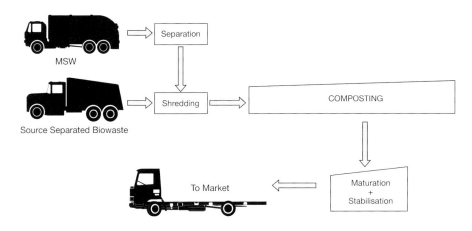

Figure 5.2. Generalised layout of a typical centralised composting site

Screening Equipment

Some form of screening equipment is routinely employed on most centralised composting plants, ranging from complex, purpose designed systems to much simpler and cruder forms. Where a site takes mixed MSW and the composting operation represents just one part of a larger waste management facility, possibly with a strong traditional landfill element, or increasingly, a recycling initiative, the screening process may be of considerable scale and importance. Even for the facility which receives source separated biowaste and therefore has no need of a major separation phase, screening may still play an important role in removing the bulk of the soil, sand, gravel and plastic bags often encountered in 'garden/yard' waste, prior to its being shredded. Aside of the obvious protection of the shredder mechanism this affords, it also goes someway towards reducing problems during the composting process itself.

Additionally, the final product may be screened as part of the finishing stages of treatment, to remove any remaining debris, unwanted residual inclusions, oversize particles and the like, thus helping produce a clean, high quality composted material, with the obvious resulting advantages for its marketing.

The type of screen to be used and details of its design, like hole size, hourly throughput and so on are, clearly, matters decided on by considerations such as input feedstock and the intended final market for the product. There are two kinds of screening devices in general use, namely trommels and vibrating screens. Trommels are common pieces of equipment throughout the wider waste management industry and form a central part of a number of MRFs designed around the removal of the biowaste fraction, often taken off in the form of 'fines', from mixed waste. Their rotating action imparts a tumbling motion to the waste, which allows trapped paper, plastic film and other unwanted or oversized items to be separated from the desired organic material. In addition, this rotary movement tends to spread the waste relatively uniformly throughout their length, thereby evening out any humps created during loading. Vibrating screens and shaking tables are popular alternatives, especially for biowaste derived compost and mulch production applications, being particularly effective in the early rejection of oversize items. For both systems there are a number of different models, produced by a variety of manufacturers to deal with the full range of operation requirements, from around 5 tonnes per hour upwards.

Process Parameters

Although the precise details of processing depend on the specific technique, equipment and management system used, several general parameters affect the efficient composting of biowaste material and the production of a useful end product.

- Temperature
- Moisture content
- Aeration
- Particle size
- Nature of the feedstock
- Processing time
- Accelerants
- Maturation and curing

Temperature

As described earlier, the temperature of the composting mass rises from ambient, reaching mesophilic (c.22 °C – c.40 °C) and thermophilic (c.40 °C – c.60 °C) ranges during different stages of treatment, this elevation of temperature resulting from the respiration and other general metabolic activities of the organisms responsible for the composting process.

This leads on to two important considerations. Firstly, it is widely accepted that in order for adequate sanitisation, the material should reach a temperature of not less than 55°C, though, as discussed in the previous chapter, there is less agreement over the required duration of this exposure. In this context, there is also the issue of windrow turning to consider, particularly for those located in cooler climates, which would intuitively seem to have a great potential for cooling when turned in the traditional *en masse* fashion. However, recent work indicates that this may not be as significant as was once thought, with average temperatures in windrows turned seven times in four weeks being 'not significantly different'[10] from those in similar piles turned only once in the same period.

Secondly, the temperature should ideally not be permitted to exceed 70°C, since above this most of the micro-organisms involved in composting die off or enter a dormant phase. This inevitably results in the decomposition slowing down or ceasing altogether, which is clearly undesirable in a commercial operation, for which such lost processing time has economic repercussions.

Moisture Content

For optimum composting, the required moisture content lies in the range of 40–70%, with a target of 60% being the most suitable. While food waste typically presents within these ideal margins, some garden waste, unless it contains large amounts of grass, can often be surprisingly dry, with a moisture content even as low as 25–30% not being unknown. This degree of dryness approaches the levels at which severe biological inhibition of the process is known to occur. To avoid this, the composting material can be roughly wetted by hoses or sprinklers, or more precisely by means of watering devices attached to the windrow turner or within the body of the in-vessel system or grinder, as appropriate. For those facilities involved in performing co-composting operations, the addition of the sewage sludge, manure or slurry itself can help to provide the necessary levels of moisture. It is equally important to ensure that the level does not rise too high, since, under these circumstances, natural aeration is reduced and there is a danger of localised anaerobic conditions developing. In addition, excessive water increases the likelihood of nutrients and potential pathogens leaching out from the material being processed. However, neither extreme need be a major problem, as a programme of careful, regular monitoring enables the moisture content to be kept within the desired range.

Evaporative losses from the surface of the composting biowaste can cause problems if left unaddressed, since water is essential to maintain the availability of nutrients to the decomposer organisms, as well as for inclusion within their own cells. It has been shown that, unlike windrow temperature, moisture is directly affected by frequency of turning[11]. Although all piles of composting matter gradually lose moisture over time, there is strong empirical evidence to suggest that moisture loss is lessened in processing regimes which turn their windrows less often.

Aeration

Composting is an aerobic process and consequently adequate aeration, or more specifically, supply of oxygen to the microbes within the material to be treated is of great importance. While home composters and the traditional gardener's heap can achieve this by natural diffusion, commercial scale operations must employ ancillary means to ensure effective aeration, as have been described earlier. During the active composting phases, measurements of mean in-pile oxygen concentration may reveal a lower level than anticipated, though this is typically a result of the high rate of gross uptake by the decomposer organisms rather than any inefficiency in aeration methodology. This can be readily verified by the fact that the same measurement made in a mature pile, characteristically reveals a return to near ambient oxygen levels, the resident biological activity and thus the oxygen demand having greatly abated.

Though windrow aeration by turning is a well-established technique, it would seem that the benefit is somewhat short lived. Though the levels do rise immediately after turning, as might be expected, the effect appears to be transient, with the oxygen concentration dropping back to the pre-turn value within hours.[12]

Particle Size

The optimum particle size for composting represents a compromise between maximising the surface area to volume ratio and minimising resistance to air flow through the matrix. Reducing the particle size exposes a larger surface area per unit volume to microbial attack, thereby enhancing both the ease and speed of biodegradation. However, there is a logical limit on this, since shredding the particles too finely enables them to pack tightly together, increasing the risk of compaction and significantly reducing inter-particle voids, thus restricting oxygen availability. Process optimisation requires a balance to be established to provide the smallest possible particle size which does not interfere with air flow. Factors of operational design may feature in this calculation, since bed depth and configuration, aeration system and input material may all exert an influence on the final ideal dimensions. For some applications, the energy cost involved in the shredding process may also need to be considered when arriving at the 'optimum' size.

Nature of the Feedstock

The nature of the biowaste to be processed, its compositional mix, the relative abundance of constituent materials and their chemical characteristics all play an important part in determining the precise details of its decomposition, though in functional terms, only the broadest of these variables are of operational concern. Of these, the most important is the carbon to nitrogen (C:N) ratio. The role of an optimal C:N ratio in avoiding nitrogen becoming a limiting factor in the overall process has been examined earlier. While it does not warrant a lengthy restatement here, suffice it to say that material high in carbon decomposes slowly. Consequently, careful management may be required to ensure that the final mix of materials entering the composting phase lies within the optimum range of values, particularly when seasonally driven inputs containing a high proportion of tree or hedge clippings occur.

Co-composting and the use of compost amendments are other allied ways of enhancing the conditions for efficient biological processing. The effect of manures and sewage sludge in increasing the moisture content has already been mentioned, but these additions also boost the available nutrient levels within the composting material, though the exact amounts of either contributed may be somewhat variable. A number of possible additives exist, either to increase the nutrient content or to act as bulking agents to improve pile structure, though for obvious reasons, ideally a steady supply of the selected amendment will be readily available, without impacting adversely on the composting facility's commercial viability. For this reason, though fertilisers and similar inorganic nutrient sources can be used, the cost of their purchase generally rules them out for use in applications treating typically 'low-value' garden or food biowastes. Their usage is more usually restricted to higher cost wastes and is particularly prevalent in the composting of contaminated soils seen in a growing number of *ex situ* bioremediation operations.

While the addition of nutrient amendments accelerates the composting rate, a measure of extra care is necessitated in their use and dependent on the particular substance to be applied, there may be attendant supplementary legal or licensing requirements with which to comply. Monitoring is of particular importance in amended or co-composting regimes, since the decomposition of the blend may proceed differently from the original material, particularly in respect of pH and this may, in turn, have an effect on the final product derived.

Processing Time

The length of time required for composting to take place depends on a number of factors, many of which have already been examined. This list includes the composting technique to be used, the input material, particle size, nutrient balance and moisture content. Processing biowaste of a garden or food waste origin can be achieved in under 3 months using an aerated, in-vessel or turned windrow system, while an un-aerated static pile may take a year or more to do the same. Much also depends on the management regime employed, since with a good basic understanding of the process variables and any inherent system limitations, it is possible to optimise conditions for maximum efficiency by attention to good operational practice. Interestingly, it has been suggested that how often a windrow is turned may have less of a relationship with the processing time than is generally supposed. Certainly, some evidence indicates that turning frequency has 'no significant effect on the amount of time required for composts to become stable' [13] with stability being defined here in terms of an oxygen uptake rate of less than 0.1mg of oxygen per gram of organic matter per hour. However, the general consensus amongst operators and the industry, though admittedly based on everyday site management rather than a specific scientific study, still appears to be that relatively frequent windrow turning is essential for the proper progression of the process.

Accelerants

For the home composter or keen gardener, a number of compost accelerants exist which speed up the onset of composting, or otherwise enhance the rate of decomposition. Generally, this is not an approach used at commercial facilities, mainly due to the scale of these operations, which would tend to make the cost of a sufficient quantity of accelerant for the large volumes of biowaste involved prohibitively expensive. However, as with the case of artificial nutrient amendments, there has been some interest in recent years regarding the possible use of accelerants for 'high-value' biowaste applications, though it remains unlikely that they will be used at sites receiving household or garden waste materials. However, some kinds of commonly co-composted substances, most notably slurries and manures, are empirically known to act as natural accelerants and it would seem that this is likely to be the only avenue for this kind of enhanced processing open for general use.

Maturation and Curing

After the active composting process itself has finished, the material is generally removed and placed in a curing pile, where it is allowed to mature without being further turned or aerated. Although this represents an additional delay before the material can be sold on and inevitably adds to the overall processing time, this stage, as has already been discussed, is of enormous importance to the stabilisation and quality of the final product.

Measuring Compost Maturity

Stability and maturity are linked aspects of compost production, which though readily and intuitively understood on the basis of largely subjective criteria of appearance and smell, can be more objectively gauged from the process monitoring regime, and particularly, by reference to certain key indicators. Fully finished compost will generally have very low levels of organic material left undecomposed, and will thus undergo little further breakdown. This is the basis for the SOUR (Specific Oxygen Uptake Rate) testing previously mentioned as a means of assessing actual microbial activity within the composting matrix. However, this may not always be the best measure since the persistence of higher levels of un-degraded organic material may be appropriate for composts expressly intended to mature *in situ*, where this final process of stabilisation after application may be of benefit. Though generally this is not the case, there are some instances for which this is of particular value. The progress of biodegradation, and hence the state of compost maturity, can be judged from routine regular measurements of the C:N ratio, or by the presence of nitrate and/or phosphate ions in the material, since composting changes these substances from their initially available organic forms, to their inorganic equivalents. The absence of ammonia in the compost is another key indicator of maturity, having being removed as the nitrification process ran its course.

Mulch Production

In the quest for a valid use for biowaste materials, one thing which it seems can sometimes be overlooked in Europe, and especially in the UK, is that there are some kinds of materials typically consigned to be 'composted' for which true composting is not the best solution. This is particularly true of the woodier materials in garden waste, which are characterised by a high C:N ratio, as illustrated in the earlier table 5.1. One solution is the 'mix and match' approach previously discussed, to produce a blend of materials with a more suitable overall balance between carbon and nitrogen. Though this unquestionably works very successfully, it is not always possible to achieve, especially where local conditions or other factors dictate an over-abundance of these high carbon materials, even if only on a temporary or seasonal basis. Moreover, for foliage and fallen leaves, or even waste bark (C:N ratios of around 60:1 and 120:1, respectively) physically accommodating and working the additional volume of 'low carbon' material required can be viewed as feasible. To do the same for hedge and tree clippings as well as unwanted timber, pallets and the like, with a typical C:N of 400:1 or more, could not be realistically attempted at most composting facilities.

For these kinds of materials, the production of mulch is a particularly useful alternative treatment, since it provides a beneficial means of re-use for the waste, while not requiring that generally scarce resources be bound up excessively. The market in the US for mulches has proved both stable and financially viable for a number of independent ventures, with allied equipment and technologies developing to keep pace with the expansion. In particular, the ability to colour the end product to remarkably consistent and precise hues has opened the way for biowaste-derived mulches to penetrate a number of landscape industry sectors which were previously dominated by other more traditional players.

The principal obvious difference between mulch and compost is that whereas compost is typically incorporated into the soil itself as an amendment to enhance soil organic matter, or used in potting mixes, mulch is laid on the surface of the soil as a landscaping feature, to reduce competition from weeds and to lessen the evaporation of soil water. While compost may increase a soil's water holding capacity, and, as will be discussed later, can help control certain soil-borne diseases, it is possible to draw a distinction between these as truly soil-integrated functions and the activity of mulch at the soil boundary level. Additionally, unlike composts, mulches generally do not add nutrients to a soil nor significantly increase its cation exchange capacity.

Although it is possible to make mulch from the same kinds of material as compost, in practice it is seldom done, with mulch production usually being reserved for the kinds of woody, 'high-carbon' wastes described. These commonly used feedstocks can themselves be divided into two broad types, based on their biodegradability and the treatment they each require varies accordingly. Pallets and other waste timber, together with hardwood and softwood logs make up one group, which can be formed into mulch simply by being chipped to the required size. After the addition of a colourant if desired, this material can then be sold either in bulk, direct to the professional end users, or bagged for domestic applications.

The second group largely consists of the woodier items of garden biowaste, like hedge and shrub prunings, which normally arrive at the treatment facility complete with their foliage attached, together with a certain amount of other general garden waste items. The production of mulch from this material follows similar lines to composting, requiring the waste to be shredded to an appropriate particle size, though this will typically be somewhat larger than for true composting, and then consigned to a windrow or pile to undergo a short period of aerobic processing. The purpose of this 'composting' phase is to allow the mulch to be rendered more stable and to destroy any weed seeds or pathogens which may be present, as the pile goes through its heating cycle. For this latter reason, as in traditional composting, the mulch should be brought to a temperature of 55°C which is then maintained for a suitable period of time, with a series of turnings being made to ensure that all of the material has been properly exposed to the heat. After the mulch has passed through this routine several times, it can be sold on much the same basis as the preceding type, though at the present time, the addition of artificial colouring to this material is not common.

In both cases, any fine material resulting during processing can be screened out and may be added either to windrows which are being used for compost production or mixed in with the final compost itself for its value as a soil amendment.

Uses of Biowaste-Derived Compost

One of the major recurrent themes of biological waste treatment is the market-limiting effect of the lack of a recognised system of standards for the product, which is discussed more fully elsewhere, and it seems likely that consumer confidence and thus market penetration will continue to be suppressed until this issue is adequately addressed. Moreover, the financial viability of biowaste processing projects cannot be wholly, or even largely, dependent on sales projections nor demand that the market supports an unachievable pricing structure. There is a clear background of interest amid both public and professional users and of all the biowaste treatment technologies, composting occupies probably the strongest position in relation to the existing tentative but expandable market, principally as a result of its 'clean' initial feedstock and consequent public image. Certainly, the analysis of compost arising from various facilities around the world would seem to indicate that acceptable

Table 5.3 Illustrative Metal Levels in Source Separated Compost Shown with the OWCA Standard (all in mg/kg)

Metal	Compost *	OWCA Standard
Cadmium	0.3	10
Chromium	22	1,000
Copper	21	400
Lead	90	250
Mercury	0.1	2
Nickel	16	100
Zinc	88	1,000

*Scarborough Organic Green Compost
Data courtesy of Scarborough Environmental Services

quality can be achieved, particularly in respect of source separated operations, as the illustrative values in table 5.3 show. Although the Organic Waste Composting Association (OWCA), and others, have attempted to establish standards, there remains an evident need for some system of universally recognised accreditation.

Aside of the obvious opportunities for compost utilisation in both domestic and civil engineering applications, there is considerable potential in the agricultural sector, though here, perhaps more than anywhere, the question of standards and public acceptability will be of paramount importance. Farmers have not been slow to learn the lessons of BSE and the furore over the possible consequences of animals reared on genetically modified feeds entering the human food chain has focused their attention ever more closely on their own supply chain issues. As discussed in the previous chapter, the final arbiters may yet turn out to be the supermarkets themselves.

Agricultural Applications

The agricultural and horticultural benefits of compost can be briefly summarised as the addition of nutrients and organic matter, together with the improvement of soil structure, which itself leads to improved aeration and a facilitation of root growth. In certain areas, it may also help mitigate the rigours of erosion and add a valuable microbial compliment to naturally poor soil. The results of various studies into biowaste-derived compost usage on agricultural land have shown that whereas high application rates generally tend to lead to relatively large increases in crop yield, at lower levels the effect is measurably less significant, being almost entirely restricted to the regulation of humus content. The ability of biowaste compost to partially substitute for proprietary chemical fertilisers has been a consistent finding in many investigations and particularly in structurally deficient soils, the use of this material can lead to better final results than would have been achieved by the application of artificial fertilisers alone. Work at the University of Kassel has also shown that compost made from food and garden biowaste fractions can have additional positive effects on both a food's quality and its storage performance. This investigation found that significantly reduced nitrates and consequently an improved nitrate to vitamin C ratio resulted in a range of produce, including cabbage, carrots, potatoes and tomatoes. Given the high cost of traditional fertilisers, the similar crop yields found to be obtained using a 50:50 co-dose of fertiliser and biowaste compost, compared with those obtainable with 100% fertiliser application[13], would suggest that savings could be made, assuming that the compost is readily available in sufficient quantity and of a quality acceptable for this use. It has been estimated that of the approximately 81 million hectares of arable land available throughout the EU, only around 3.7% would be required to be made available for the application of compost, at a rate of ten tonnes per hectare per year, even if the maximum possible quantities of biowaste compost were generated. Presently, only around 13.5% of this estimated total is actually produced[14].

Plant Disease Suppression

Another area which has generated considerable interest is the use of biowaste composts in plant disease suppression, since these can cause some greenhouse, nursery and vegetable crops to suffer very expensive losses. Until about 1930, the traditional means of protection lay in a programme of crop rotation coupled with the use of animal manures and green mulches. Since then, many growers have followed a regime of chemical fumigation, usually involving methyl bromide, to combat soil-borne pathogens, which tend to accumulate, particularly under intensive horticultural monoculture systems. This fumigant also destroys weeds in the soil and resident insect pests, making it widespread and popular and for a considerable number of operations, commercial viability is directly linked to its use. However, methyl bromide kills microbes and insects indiscriminately and, having been implicated in ozone depletion, is scheduled to be phased out by 2005 under the terms of the Montreal Protocol. As a result of this and other concerns regarding residual bromine in food and ground water, it has already been banned in Germany and Switzerland and has also been phased out in the Netherlands, once Europe's largest user of methyl bromide soil fumigation. Consequently, finding an alternative means of crop-specific disease control has become something of a priority for the industry and the possible use of biowaste compost and compost extracts is being explored alongside soil steam pasteurisation, ultraviolet treatment and the development of resistant cultivars both by selective breeding and genetic modification.

Much of the original work in this area has tended to concentrate on compost extracts, also often termed 'compost teas' or occasionally 'compost watery extracts', and specifically their ability to protect crops against foliar diseases, and to act as inoculants, restoring or enhancing sub-optimal soil micro-flora. Many research projects around the world, most notably from Germany, Israel, Japan, the UK and the US, have produced evidence of the effectiveness of compost extracts in the control and suppression of a number of diseases through natural means, leading to a reduced requirement for artificial agro-chemicals. Selected results from these studies are shown in table 5.4.

The biological control of plant pathogens by these extracts takes place as a result of their action on the leaf surface and associated circum-phyllospheric microorganisms. A variety of mechanisms are thought to contribute towards the overall disease suppression effect, including direct competition with the pathogens, induced disease resistance, and the inhibition of spore germination. In addition to the bacteria, yeast and fungi already established as active components, a number of chemicals, especially phenols and various amino acids also play a role in disease control. However, the major effect would appear to be biological, since fine filtration and sterilization by heat treatment significantly reduce extract efficacy[23].

There are two main methods for the preparation of compost teas, known as 'fermented' or 'aerated' extraction. The former was the original approach, developed by German researchers, and involves mixing around six or eight unit volumes of water per unit volume of compost and permitting them to 'ferment' for a given period, usually between three days and a week, before being strained for use.

The alternative method arose from work done in the US and Austria which favoured the aeration of the compost tea as it was forming, typically passing the liquor

Table 5.4 Compost Extracts Indicated in the Suppression of Plant Diseases

Plant Disease	Compost Extract
Botrytis cinerea Grey Mould of beans & strawberries	Cattle Compost Extract[15, 16]
Fusarium oxysporum Fusarium Wilt	Bark Compost Extract[17]
Phytophthora infestans Potato Blight	Horse Compost Extract[18]
Plasmopara viticola Downy Mildew of grapes	Manure-Straw Compost Extract[19]
Sphaerotheca fuliginea Powdery Mildew of Cucumbers	Manure-Straw Compost Extract[20]
Uncinula necator Powdery Mildew of grapes	Manure-Straw Compost Extract[21]
Venturia conidia Apple Scab	Spent Mushroom Compost Extract[22]

through a compost containing vessel, collecting it and then establishing a cycle involving multiple recirculations, which both concentrates and aerates the resulting extract. The major advantage of this more active approach is that the tea is ready for use much sooner than under the previous method, and often in less than ten hours.

The finished extract can be sprayed on young transplants or seedlings as a foliar drench, in some instances at rates of around a hundred gallons per acre, or a thousand litres per hectare. In this context it is important to emphasise the importance of only using mature, stabilised material for the production of teas, since many of the unwelcome problems of improperly finished composts apply equally, if not more so, to this application.

It has also been established that many properly prepared biowaste composts themselves have the ability to suppress and control soil-borne plant diseases, especially where mature compost is mixed with soil or other growth media. Incorporation of this kind with soil known to contain the relevant plant pathogens has been reported as effective against a number of diseases, including those caused by various species of *Phythium*, *Phytophthora* and *Fusarium* as well as *Rhizoctonia solani*, which is a major threat to many kinds of plants, particularly when young. Investigations into the mechanisms of action really first began when it was widely noticed by the horticultural industry that composted tree bark appeared to suppress root rots caused by *Phytophthora* spp. Amongst some of the most important members of the soil community are the small fungi which typically surround plant roots, known as vasicular arbuscular mycorrhizae (or VAM) and which are intimately involved in plant water and nutrient uptake. The original work on *Phytophthora*-caused disease led on to the discovery that woody plants in general, which require mycorrhizae for growth, also thrive better in soil/bark compost mixes than in methyl bromide-treated soil alone, which in turn stimulated further research. Current understanding of the processes suggest that, unsurprisingly, the pathogen inhibition of biowaste derived compost is due in the main to the indigenous micro-flora of the mature material and several additional groups of bacteria, yeast and fungi have also been identified as the principal active agents in this respect. This natural biological control property can, however, be diminished if the treatment the compost receives after maturation alters the resident microbial community, or gives rise to conditions which hamper their action. This has been shown in laboratory trials, in which biowaste compost was deliberately exposed to high temperatures or microwaves in order to disturb the micro-organism balance. This resulted in disease suppression becoming either inhibited or lost entirely and, in some cases, led to 100% plant mortality[24]. The finding that excessive heat significantly changes compost properties has very important potential ramifications for the common practice of storing the finished material outdoors, often in full sun, particularly if it has been packed into plastic bags for sale.

Environmental Applications

The possibility of using compost in erosion control has already been touched on. This application of biowaste derived material is a relatively new one, which appears to hold much promise. Studies have shown that compost can generally equal the results of more traditional methods of bank stabilisation for slopes of up to $22\frac{1}{2}°$, when layered on at a depth of around 10cm (4 inches). Compost blends, consisting of around 20% sand, gravel or chipped bark, thoroughly mixed into coarse biowaste compost, can often outperform the established techniques. The thick layer of surface dressing absorbs much of the energy of incident rainfall, which is one of the main initial causes of fine soil movement, soaking up much of the volume of water, thereby reducing flow velocity and consequent soil losses by run-off. When the conditions are drier, the heavy blanket layer of relatively larger, coarser particles protects the soil itself from becoming wind-blown. In practical applications of this technique, coarser mulches and composts, and particularly those containing a high proportion of woody material, are not seeded in the first growing season. However, since they tend to naturalise *in situ*, grassing over the bank, to make use of both the resulting cosmetic improvement and the additional soil holding properties of its roots, becomes possible in subsequent years. For major landscaping projects associated with site remediation or civil engineering developments, this application may have particular relevance for the less 'clean' kinds of compost produced from mixed MSW operations, though probably only for those sites situated away from residential areas.

The widespread application of higher quality compost has also been identified as one of the possible ways to combat both erosion and the depletion of organic matter from soils throughout the Mediterranean region.[25] In particular, soil quality in Italy, Greece and the Iberian peninsular is reported[26] as rapidly dropping, largely due to mineralisation and the insufficient return of humus material. Consequently, a sizeable potential outlet for the EU's composted biowaste exists in this region. If realised, this could significantly reduce the amount of European arable land required to accommodate EU's potential biowaste compost production mentioned earlier, since the functional deficiency of organic material in these soils would indicate a higher application rate per hectare than the annual ten tonnes per hectare which formed the basis for the original calculation.

The use of compost as a large scale amendment has other advantages which has led to its consideration for wider revegetation and reclamation projects. Viewed in the larger context of sustainability, particularly in respect of post-industrial scenarios, this use of community biowaste has a clear appeal. In research using relatively dry garden-waste derived material, it has been shown not only to lessen soil wash-off, but also to have the ability to filter and bind contaminants in the stormwater run-off itself[27], thereby making a significant reduction in localised soil pollution. Where remediation is to be carried out beside other sites, either still in current use, or awaiting their own clean-up, a biowaste compost layer may well offer additional protection at the critical early stages of establishment.

The high organic matter and intrinsic biological activity of such material makes it ideally suited to a number of types of reclamation projects, especially where marginal or low quality soils are involved. Aside of erosion control, the improvement of soil quality and the addition of humus and a thriving microbial community, which have already been discussed, compost also offers enhanced plant establishment and the potential immobilisation of certain kinds of toxic substances, most notably some metals. It has been shown, for instance, that the addition of composted material to city soils containing lead in concentrations of 1600 parts per million, can result in up to a 60% decrease in the metal's bio-availability. Consequently, applied at rates varying between 50 and 500 tonnes per hectare, composts have been used in the reclamation and revegetation of a number of different kinds of sites, including finished landfills, old mine workings, roadsides and disused factories. Some of the most interesting current trials of this kind are examining the use of composted material in a treatment–train approach to recondition soil which has been subjected to destructive remediation techniques, like thermal processing. This form of remediation, though very effective at destroying many unpleasant and otherwise recalcitrant chemical pollutants, also destroys soil structure, leaving a final product with no organic matter or resident organisms, and of a severely limited cation exchange capacity. The use of biowaste derived material to redress this seems a particularly appropriate use, not least since the same political will and consequent financial forces which are set to support the diversion of biowaste from landfill also play a significant role in the growth of land remediation.

Field research throughout the UK has demonstrated increased water holding capacity as another significant benefit to be gained from the large-scale recycling of composted biowaste material into soil. With the addition of around 250 tonnes of compost, each hectare treated has been found to be able to hold between 1000 and 2500 tonnes of rainwater[28]. The value of this is, perhaps, most clearly evident in those trials conducted in the loose, sandy soils of East Anglia, where even in these wind-prone conditions, at the 2000 tonnes of water per hectare level, in most years crops can be grown without any further irrigation need to be met[29]. Even in seasons of high sunshine and low rainfall, or for those crops which have particularly high requirements, the demand is greatly reduced, which is scarcely surprising given the fact that composts, particularly those which are applied 'young' and allowed to finish their maturation *in situ*, can absorb and retain between two and ten times their own weight in water[30]. This leads not only to the stated lower requirement for supplementary watering, but also reduces drought-stress in the plants and reduces nitrate leaching.

The combination of high organic content coupled with this ability to retain large amounts of water has led to the use of compost in wetland construction, especially in the US, where federal environmental regulations require the reestablishment of this type of habitat as a means of improving water quality. Since the express intention of this is to provide a wetland which behaves like a natural system in terms of both its hydrology and biology, a humus-rich, microbiologically active medium closely replicating the physical and chemical properties of native soils is required. It has been found that biowaste derived composts make excellent components of manufactured wetland soils, with vegetation often becoming established on such sites more quickly than usual.[31]

It is almost certain that the major bulk users of compost hold the key to large-scale product utilisation, and marketing strategies will need to be adjusted to reflect their particular needs. All of the land professional sectors have their own specific performance criteria, which requires compost producers to take account of a number of factors in promoting their material. Aspects like the compost's ease of application, growth promotion, moisture retention properties, nutrient content, potential for soil structural improvement and its ability to be handled by existing equipment, may all be important considerations in opening up these specialist markets.

Health Issues

Any consideration of composting as a means of biological waste treatment must take into account the question of health risks, both to site workers and the general public. Most of these issues are common to all forms of biowaste processing, and, indeed, many are applicable to the whole of waste management. These more general concerns have already been discussed more fully in the previous chapter. However, in common with biowaste collection and MRFs, composting itself, together with the household storage of biowaste for composting, raise certain specific additional matters which need to be separately addressed.

The most important of these is the issue of bioaerosols, a blanket term employed to describe a whole range of airborne microbiological contaminants, including live bacteria and fungi, their spores, whole dead micro-organisms, their component structural parts and certain biochemical inclusions. As was discussed previously, the site operatives themselves are the most likely group to be affected and good preventative measures have been widely suggested as the best way of dealing with the risk. However, most of the concern has tended to concentrate on fungal spores, while the potential extent of other agents has only recently begun to be more fully appreciated. Of these, certain cell wall components, particularly endotoxin from Gram-negative bacteria and glucan from fungi, have been shown to have a number of toxic properties and can cause respiratory sensitisation and a wide ranging group of systemic symptoms like headaches, fever, tiredness and joint pain, known as Organic Dust Toxic Syndrome.[32] Although the level of risk posed by bioaerosols in waste management forms part of ongoing research, a number of European universities having been funded for collaborative studies under the BIO-MED2 programme, EU occupational exposure standards are yet to be established. A number of Member States, notably Germany, Denmark and Holland, have addressed the issue individually, based on the known health effects and putative thresholds found by various studies into the matter. Endotoxin at the 7 mg/m^3 level has been associated with sub-clinical effects in the immune system of waste collectors and 10 mg/m^3 of glucan is believed to be causal in a variety of physical effects, including respiratory inflammation, particularly in those with pre-existing asthma or similar sensitivity conditions[33]. Initial results of research into the storage of biowaste for composting has shown that, dependent on the type of bag or container in which the material is kept, the levels of these contaminants may be as high as 250 mg/m^3 or more. Since earlier investigations have shown levels of viable micro-organisms at some UK MRFs sufficient to cause sensitisation in some individuals[34], it seems likely that the situation for workers at such facilities separating biological waste for composting is a valid cause for concern.

In many ways, potential risks arising from the storage and separation of biowaste are common to all forms of biological waste treatment, but there are reasons to view compost sites particularly carefully in respect of these bioaerosol contaminants. Although these facilities have not yet themselves been examined in detail, there is evidence of a range of endotoxin levels from 5–100 mg/m^3 and a history of acute and chronic non-immune allergic respiratory inflammation amongst the site workers[35]. In the context of composting as a universally acceptable biowaste management tool, this issue will inevitably need to be properly addressed.

Suitability for MSW

The main requirements of any biological waste treatment are that it should be able to process the particular type of material available, resulting in a bulk reduction, the creation of a useful, and ideally revenue-generating, product, efficiently and at a realistically affordable cost. On this basis, composting clearly qualifies. The relatively modest cost of establishing sites using simple composting techniques is a major attraction, particularly when compared with the higher capital cost requirements of certain other approaches. Also, the technology itself is more widely appreciated and understood, giving it less of a 'high-tech' mystique, although this can, of course, be something of a mixed blessing. Where composting can score heavily is with its in-built ability for expansion, particularly in respect of windrow or static pile type operations. Unlike bioreactor-driven systems, the scope of which can only be extended by the modular addition of extra units, simple composting can be increased incrementally, merely requiring sufficient land onto which more biowaste may be laid. This does not, clearly, apply to rotary drum or in-vessel systems. However, for the majority of local authorities contemplating biowaste treatment on a municipal scale, incremental expansion is a very real factor in explaining why windrows or static piles represent the most commonly adopted approaches.

Though the marketing of all biowaste products is largely in its infancy, and the issue of standards remains a significant obstacle to current expansion, there are many examples throughout the world of such materials, having been specifically produced by composting, being successfully sold. The contribution of the product has sometimes either been over-sold or assumed too great an influence on the commercial viability of proposed plants. It is unrealistic to assign too high a price for this kind of material.

Biowaste compost, particularly sold in bulk, will not command the same level of premium available for small bags of specialist proprietary brands sold at garden centres, especially against a backdrop of uncertain product quality. No commodity can succeed in the market-place if it does not offer the customer some advantage, be that concrete and financial, or simply perceived. The 'green' image is a powerful plus, but it is not enough on its own. The facilities which perform best seem to be those which have established a simple approach based on a sound view of the opportunities in the local area, a quality product and a balance between production cost/revenue stream cycle. The compost itself can then be looked on as *at least* cash-neutral, covering its own production costs, or *ideally*, cash-generative, with other factors contributing to the overall project viability, like gate-fee charges on 'trade' biowaste accepted from landscape professionals and even transfer credited savings against the disposal charges from other departments in the same municipality.

Composting has been the first choice for many authorities around the world when faced with the need to do something other than simply bury or burn their waste. Accordingly, this has lent a certain aura of familiarity and respectability to composting, which makes it, perhaps, more readily considered than other technologies, which may be viewed with more suspicion. The issues of capital and running costs, coupled with the perceived marketability of the product and the inherent simplicity of the process itself make composting an un-intimidating approach to contemplate and a typically popular potential solution amongst the public. With many councils having to begin to consider their positions in respect of biological waste treatment, the flexibility of scale and the possibility of smooth expansion offered by windrow systems in particular, make composting an attractive way for them to ease themselves in. As with all forms of waste management, there are factors which must be taken into account, some local, others economic or political and most of the remainder, operational. However, composting has a record of success for a wide variety of tonnages, over a range of contract terms and under differing local conditions, which make it a clear front runner in any discussion of biological waste treatment.

References

1. Bess, Vicki, *Evaluating the Microbiology of Compost*, BioCycle magazine, May 1999, pp. 62–63.
2. 'Off-the-record' comments to the author.
3. Wright, Mick, *Home Composting: Real Waste Minimisation or Just Feel Good Factor?* Wastes Management, The Journal of the Institute of Wastes Management, September 1998, pp. 27–28.
4. Ibid
5. Ibid
6. *Backyard Composting in New York City; A Comprehensive Program Evaluation*, (June 1999) New York City Department of Sanitation (www.ci.nyc.ny.us/).
7. Wright, Mick, *Home Composting: Real Waste Minimisation or Just Feel Good Factor?*
8. Ibid
9. *A Way with Waste, A Draft Waste Strategy for England and Wales* (Part 2), Department of the Environment, Transport and the Regions, June, 1999, p. 90, section 7.68.

10. Michel, F.C., Forney, L.J., Huang, A.J.F., Drew, S., Czuprenski, M., Lindeberg, J.D. and Adinarayana Reddy, C., *Effects of Turning Frequency, Leaves to Grass Mix Ratio and Windrow vs Pile Configuration on the Composting of Yard Trimmings*, Michigan State University Web Site.
11. Ibid
12. Ibid
13. DHV Environment and Infrastructure BV (Amersfoort), in cooperation with Plancenter Ltd, (Helsinki) and University for Soil Management, (Vienna) *Composting in the European Union*, a final report to the European Commission DGXI, Environment, Nuclear Safety and Civil Protection, 1997, p. 33, discussing results of work jointly done by USDA and the Pakistan Agricultural Research Council and independently confirmed by the University of London.
14. Ibid, p 28.
15. Weltzein, H.C. *The Use of Composted Materials for Leaf Disease Suppression in Field Crops*. Crop Protection in Organic and Low-Input Agriculture. BCPC Monographs No. 45, British Crop Protection Council, 1990 p. 115–120.
16. Elad, Y., and Shtienberg, D. *Effect of Compost Water Extracts on Grey Mould (Botrytis cinerea)*. Crop Protection. Vol. 13, No. 2. 1994, p. 109–114.
17. Kai, Hideaki, Tohru Ueda, and Masahiro Sakaguchi. *Antimicrobial Activity of Bark-compost Extracts*. Soil Biol. Biochem. Vol. 22, No. 7. 1990, p. 983–986.
18. Weltzein, H.C. *The Use of Composted Materials for Leaf Disease Suppression in Field Crops*.
19. Weltzein, H. C. *Some Effects of Composted Organic Materials on Plant Health*. Agriculture, Ecosystems and Environment. Vol. 27. 1989 p. 439–446.
20. ibid
21. ibid
22. Cronin, M.J., Yohalem, D.S., Harris, R.F. and Andrews, J.H. *Putative Mechanism and Dynamics of Inhibition of the Apple Scab Pathogen Venturia inequalis by Compost Extracts*. Soil Biology & Biochemistry. Vol. 28, No. 9. 1996, p. 1241–1249.
23. Trankner, Andreas. *Use of Agricultural and Municipal Organic Wastes to Develop Suppressiveness to Plant Pathogens*. Biological Control of Plant Diseases: Progress and Challenges for the Future. Edited by E.C. Tjamos, G.C. Papavizas, and R.J. Cook. NATO ASI Series No. 230. Plenum Press, New York, NY. 1992. p. 35–42.
24. E. W. Richardson, personal comment.
25. DHV Environment and Infrastructure BV (Amersfoort), in cooperation with Plancenter Ltd, (Helsinki) and University for Soil Management, (Vienna) *Composting in the European Union*, a final report to the European Commission DGXI, Environment, Nuclear Safety and Civil Protection, 1997, p. 27.
26. Ibid
27. Alexander, R., *Compost Markets Grow with Environmental Applications*, BioCycle Magazine, April, 1999, p. 43.
28. Butterworth, W., *A Top Idea That Holds Water*, Wet News, (Water and Effluent Treatment News), Volume 5 Issue 17, October 1999, p. 4.
29. Ibid
30. Ibid
31. Alexander, R., *Compost Markets Grow with Environmental Applications*, BioCycle Magazine, April, 1999, p. 48.
32. Gladding, T., *Airborne Microbiological Contaminants: Health and Safety Concerns in Waste Management*, Wastes Management, The Journal of the Institute of Wastes Management, January 1999, p. 29.

33. Ibid
34. Ibid
35. Douwes, J., Dubbeld, H., Van Zwieten, J., Woutes, I., Doekes, G., Heederik, D. and Steerenberg, P, *Work Related Acute and (Sub-) Chronic Airways Inflammation Assessed by Nasal Lavage in Compost Workers,* Agricultural Environmental Medicine, Volume 4, 1997, p. 149–151.

CHAPTER 6
Anaerobic Digestion (AD)

Although it is seen in some circles as something of a new technology, there is nothing especially novel about anaerobic digestion. Being particularly suited to wet biowaste, it has been widely used in the water industry to treat sewage since the end of the last century, and gained good recognition in certain countries, notably Denmark and Germany, for the treatment of agricultural and municipal biowastes. However, this method still has to establish a track record for the treatment of MSW-derived biowaste in the UK and elsewhere, and while it is receiving increasing interest, individual national concerns as to efficacy of direct AD technology transfer, inevitably slow its acceptance and wholesale adoption.

Anaerobic digestion is one form of the naturally occurring processes of decomposition and decay, by which organic matter (animal waste or biomass) is broken down to its simpler chemical constituents. In nature, this type of breakdown is typically associated with wet, warm and dark environments, as characterised by the digestive tract of ruminant herbivores or the bottom ooze of eutrophic lakes. Applied to waste management, AD requires the close, artificial replication of such conditions, within a digester or bioreactor cell, where the ingress of oxygen is prohibited, the contents are mixed and temperature, acidity and other internal environmental factors are optimised for bacterial action. The following generalised process flow chart gives an overview of MSW treatment by this method, though there are a variety of different operating regimes and digester designs, which may vary the specifics of the procedure in one way or another.

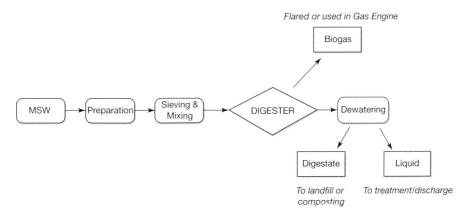

Figure 6.1. Generalised process flow chart

An Overview of the Main Processes of Anaerobic Digestion

In the optimised internal environment of a digester, large organic molecules are converted, chiefly, into methane and carbon dioxide by the action of bacteria. Under ideal conditions and in the complete absence of free oxygen, this ultimately proceeds to yield fully reduced methane CH_4 and fully oxidised carbon dioxide CO_2.

The reality of this breakdown at the microscopic level is chemically very complex, involving hundreds of potential intermediary reactions and compounds. In addition, many of these each have further need of specific synergistic chemicals, catalysts or enzymes.

In very general terms however, it is possible to simplify the overall biochemical reaction to:

$$\text{Organic material} \longrightarrow CH_4 + CO_2 + H_2 + NH_3 + H_2S$$

It must also be borne in mind that some organic materials, for example, lignin, effectively do not digest, nor, obviously, do non-organic inclusions within the waste.

It is widely considered that there are three effective temperature ranges for anaerobic digestion, each of which has its own favoured group of bacteria and its own set of characteristic advantages and disadvantages. These ranges are:

Cryophilic ($< 20°C$)
Mesophilic ($20–45°C$)
Thermophilic ($>45°C$)

The Three Stages of Anaerobic Digestion and Methanisation

There are three main stages to anaerobic digestion:

1. Hydrolysis
2. Acidogenesis
3. Methanogenesis

Hydrolysis

During hydrolysis, complex insoluble organic polymers, such as carbohydrates, cellulose, proteins and fats, are broken down and liquefied by the extracellular enzymes produced by hydrolytic bacteria. This makes them more easily available for use by the acidogenic bacteria of the next stage. In general, proteins present in the waste are converted into amino acids, fats into long-chain fatty acids and carbohydrates into simple sugars. The liquefaction of complex compounds, and especially cellulose, to simple, soluble substances is often the rate limiting step in digestion, since bacterial action at this stage proceeds more slowly than in either of the following. The rate at which hydrolysis takes place is governed by substrate availability, bacterial population density, temperature and pH.

Acidogenesis

Acidogenesis, sometimes split into *acidogenesis* and *acetogenesis*, is characterised by the production of acetic acid from the monomers released in the preceding stage and volatile fatty acids (VFAs) which are derived from the protein, fat and carbohydrate components of the feedstock. The main products of this stage are acetic, lactic and proprionic acids and the pH falls as the levels of these compounds increase. Carbon dioxide and hydrogen are also evolved as a result of the catabolism of carbohydrates, with the additional potential for the production of methanol and/or other simple alcohols. The proportion of the different by-products produced depends on the environmental conditions, to some extent, and, more largely, on the particular bacteria species present.

Methanogenesis

Methanogenesis involves the production of methane from the raw materials produced in the previous stage. This is brought about by obligate anaerobes, whose growth rate is, overall, slower than the bacteria responsible for the preceding stages.

The methane is, then, produced from a number of simple substrates: acetic acid, methanol or carbon dioxide and hydrogen. Of these, acetic acid and the closely related acetate are the most important, since around 75% of the methane produced is thus derived, according to the equation below:

$$CH_3COOH \longrightarrow CH_4 + CO_2$$

Methane-forming bacteria may also use methanol:

$$CH_3OH + H_2 \longrightarrow CH_4 + H_2O$$

or carbon dioxide and hydrogen:

$$CO_2 + 4H_2 \longrightarrow CH_4 + 2H_2O$$

There are other potential substrates for methane-producing bacteria, such as formic acid, but they are beyond the scope of this discussion, since they do not routinely occur within the anaerobic digestion of municipal (or similar) solid wastes.

Aside of the fact that the production of methane yields a useful fuel source, the actions of the associated methanogenic bacteria play a vital role in maintaining the wider breakdown process. By converting volatile fatty acids (VFAs) into methane and associated gases, any trend towards an increase in VFA concentration (and thus a decrease in pH) is reduced. In this way, the acid/base equilibrium is naturally regulated, at least in part, and the attendant potential for biochemical inhibition and/or bacterial population destruction provided by the acidification of the digester environment is largely removed. Methanogens are pH sensitive, the required range being mildly acidic (6.6–7.0) and problems are likely to be encountered if the pH falls much below 6.4.

Moreover, should this stage not progress properly, the required stabilisation of the waste will not be achieved and the volatile fatty acids produced prior to the methanogenesis phase will have serious implications with regard to the final use or disposal of the material derived.

The Bacterial Ecology of Anaerobic Digestion

There are known to be four main groups of bacteria involved in the stages of AD:

I   Hydrolytic fermentative bacteria (e.g. *Clostridium*, *Eubacterium* and *Peptococcus*)
II Acetogenic bacteria (e.g. *Desulfovibrio*, *Syntrophobacter* and *Syntrophomonas*)
III Acetoclastic methanogens (e.g. *Methanosarcina* and *Methanothrix*)
IV Hydrogenotrophic methanogens (e.g. *Methanobacterium* and *Methanobrevibacterium*)

The relationship of these groups to the phases of anaerobic digestion outlined can be simply shown as below:

However, it will be readily apparent that this does not reflect the entire complexity of the process of substrate dissimilation under anaerobic digestion, which would be better represented by the diagram below (though remembering that this, too, is an illustrative simplification).

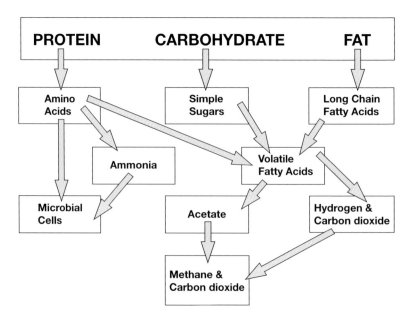

Figure 6.2. Simplified diagram of substrate dissimilation

Methanogenic bacteria, together with extreme halophilic and thermoacidophilic bacteria, make up a group of micro-organisms which forms the distinct biological kingdom *Archaebacteria*. Methanogens are strict obligate anaerobes, being killed by exposure to even relatively small amounts of oxygen, though not requiring the absolute absence sometimes suggested.

Methanogens are a morphologically diverse group, consisting of 12 genera:

Methanobacterium
Methanobrevibacterium
Methanomicrobium
Methanospirillum
Methanothrix
Methanococcus
Methanococoides
Methanogenium
Methanosarcina
Methanothermus
Methanolobus
Methanoplanus

Originally these bacteria were classified in the time-honoured traditions of taxonomy, according to their morphology, thereby spreading them through more widely known groups. However, they form a physiologically coherent grouping, based on their methanogenic metabolic abilities and more recent phylogenetic analysis based on sequence characterisation of ribosomal ribonucleic acid (rRNA) has caused their reclassification into the kingdom *Archaebacteria*. Additionally, the biochemistry of methanogens share certain older features with others of the *Archaebacteria*; ether-linked proty-isoprenoid glycerol lipids and the absence of thymidine in the so-called 'common arm' of transfer ribonucleic acid (tRNA).

Although the biochemical processes of hydrolysis, acidogenesis, acetogenesis and methanogenesis, and the bacteria responsible for effecting them, are the primary activity desired within a digester, other bacterial types and breakdown pathways also come into play. For example, homoacetogens can ferment a number of substrates, including formate, carbon dioxide and hydrogen, into acetate, which leads them into competition with methanogens for the same substrate. The extent of this competition is difficult to assess and thus, not yet completely understood. In addition, under certain conditions and provided with the right substrate, denitrifying bacteria and sulphate-reducing bacteria can readily out compete digester resident methanogens.

Within the digester, also, however, there exists extensive interspecies co-operation, a clear example of which is the manner in which the four main groups of bacteria interact to effect the three stages of AD. A further important aspect of such co-operation will be discussed later, relating to the role of hydrogen in regulating methane formation.

The Chemistry and Production of Biogas

As has been mentioned previously, there are two main pathways for methane production under AD. These can be shown in a simplified diagrammatic form as follows:

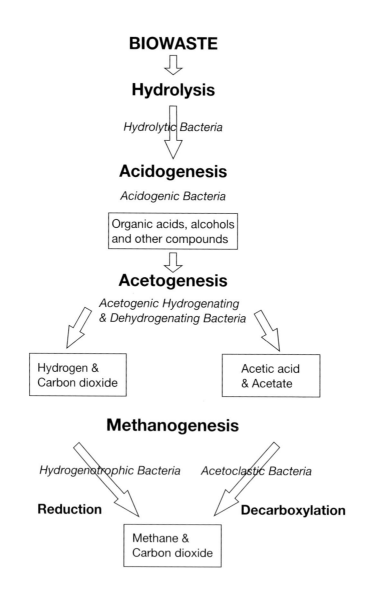

Figure 6.3. The methanisation of biowaste

Anaerobic Digestion (AD)

Table 6.1 Typical Biogas Quality [1]

	% by volume
Methane	63.8
Carbon Dioxide	33.6
Oxygen	0.16
Nitrogen	2.4
Hydrogen	0.05
Carbon Monoxide	0.001
Saturated Hydrocarbons	0.005
Unsaturated Hydrocarbons	0.009
Halogenated Compounds	0.00002
Hydrogen Sulphide	0.00002
Organosulphur Compounds	0.00001
Alcohols	0.00001
Others	0.00005
Water	0.001–0.004

Recent work[2] has identified a more complete list of trace gases within the 'others' category above, though their presence in any given sample of produced biogas will obviously depend on the nature of the biowaste undergoing digestion:

1,2 Dichloroethene	Alkylbenzene
Butylcyclohexane	Carbon disulphide
Propylcyclohexane	Methanethiol
Decane	Dichlorobenzene
Undecane	Ethylbenzene
Dodecane	Trimethylbenzene
Tridecane	Toluene
Dimethyl disulphide	Nonane
Sulphur dioxide	

Table 6.2 Major Physical Properties

Methane explosive range	5–15% vol
Hydrogen explosive range	4–74% vol
Methane density	0.72 kg/m^3 @ 20°C
Hydrogen density	0.09 kg/m^3 @ 20°C
Carbon dioxide density	1.97 kg/m^3 @ 20°C
Calorific value	5.5–6.5 kWh/m^3

A number of authorities have sought to establish models for the prediction of biogas production, ranging from the very simplistic to the highly sophisticated. Generally, it is widely accepted that the gas evolution/cellulose decomposition curves can be characterised as having five main phases, of which the main points relevant to the present discussion are:

Phase I
Maximum cellulose loadings; oxygen content drops to near zero; nitrogen, and carbon dioxide at atmospheric levels (20% and 78% respectively).

Phase II
Carbon dioxide, hydrogen and free fatty acids levels rise to peak values; nitrogen levels fall to around 10%; cellulose breakdown begins.

Phase III
Carbon dioxide decreases to plateau at around 40%; methane production commences and achieves plateau at around 60%; free fatty acids decrease to hold at minimal levels; cellulose breakdown continues at a linear rate with respect to time; nitrogen falls to near zero.

Phase IV
Plateau phase with carbon dioxide at c.40% methane at c.60% and free fatty acids at less than 5%; cellulose declines steadily throughout this phase.

Phase V
Cellulose becomes fully decomposed, resulting in the tail off to zero of methane and carbon dioxide; oxygen and nitrogen regain atmospheric levels (20% and 78% respectively).

These are best represented in graphic form:

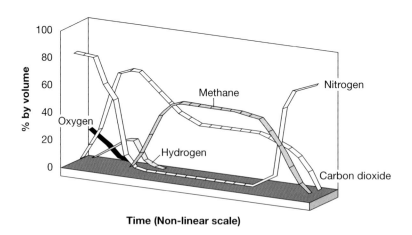

Figure 6.4. Biogas generation

The Role of Hydrogen

As can be readily appreciated from the preceding discussion, there exists an obligate, syntrophic relationship between the hydrogen-producing acetogenic bacteria and the hydrogen-utilising methanogens, the basis of which involves interspecies hydrogen transfer. In essence, to produce the required thermodynamic conditions within the digester favourable to the conversion of higher volatile fatty acids and alcohols to acetate, a low hydrogen partial pressure must be maintained. This, clearly, requires an active population of hydrogen-utilising methanogens, which ensures that the alcohol and acid degradation are not inhibited. In the event of hydrogen partial pressure rising beyond around 10^{-4} atmospheres, acetogen catabolism proceeds to produce butyric, lactic, proprionic and other acids in preference to acetic.

In this way, hydrogen may be said to have a regulatory role in methane production, and in the control of the redox potential, since, in extremis, elevated hydrogen levels lead, inevitably to the accumulation of higher volatile fatty acids beyond the self-buffering point, hence causing significant lowering of the pH within the system. This, of course, begins a downward spiral effect, as the methanogens themselves are inhibited by increased acidity, leading, if left unchecked, to the cessation of methane production and process collapse.

Process Variables and Operational Conditions

In a commercially operational AD facility, there is an obvious requirement to achieve a significant reduction in the biowaste feedstock within a reasonable time, for both practical and business reasons. This clearly places a high priority on the attainment of an optimised internal environment, wherein all the necessary biological activity and biochemical reactions can proceed as swiftly and efficiently as possible. There are a number of variables with the potential to influence the anaerobic digestion of biowaste, which can be broadly divided into physical and chemical factors.

Physical

Digester Mixing

Efficient mixing of the digester contents has two principal effects. Firstly, it tends to homogenise the digesting material, thereby avoiding any localised concentrations of any given substance, dead zones or scum formation. Secondly, it improves contact between the material itself and the digester's resident bacteria, thereby increasing their ease of access to the available nutrients and facilitating the desired breakdown of the biowaste feedstock.

Mixing also has a number of additional benefits to digester operation. The agitation assists the removal and dispersal of by-products from the micro-organisms, which is of obvious advantage in a system which relies on a series of different bacterial types, each producing the required substrate for the next. Though it has been suggested that mixing has little effect on the rate of methane production or its final quality, others have found that it acts to enhance the methane production, by removing the gas from the immediate zone of generation, as it forms. It seems likely, then, that this has more to do with other aspects of digester design or operational practice and may well depend entirely on the type of system or regime in use.

Finally, the physical mixing of the contents helps to maintain a more uniform temperature within the digester.

Temperature

Of the three temperature ranges mentioned earlier in this chapter, only mesophilic and thermophilic feature in operational AD systems. Generally speaking, this means digesters run at around 35°C and 55°C respectively, though, in absolute terms, the optimum temperature may depend on the exact type of biowaste material undergoing treatment and the digester used. While the relative advantages and disadvantages of systems run at each of these temperature ranges will be discussed later, suffice it to say at this point that any anaerobic digestion process requires a relatively constant temperature to progress at its greatest efficiency.

Retention Period

The biowaste needs to be allowed to remain in the digester until the required level of treatment has been achieved. Though the actual amount of biowaste broken down depends on its own character and ease of biodegradation, the availability of bacteria and time, temperature governs both the rate of breakdown itself and the particular species represented in the bacterial population. However, while there exists a direct relationship between temperature and retention period, it is also the case that in AD systems which split the digestion process into two or more separate stages, the retention period is shorter than would be the case for an otherwise comparable digester.

Such separation of the acidogenic and methanogenic stages into a distinct two-phase process allows for the optimisation of both, at least in theory. At the laboratory scale, this has been demonstrated using a mixed feedstock of MSW-derived biowaste and sewage sludge in two-step, completely mixed digester, with the requisite phase-isolation being achieved by the careful manipulation of the pH of the reactor environment. However, this is an approach which has never really found widespread favour with commercial operators of AD systems, though those who have used it claim greater efficiency, a higher biogas yield and enhanced process stability. However, the capital costs involved are obviously higher as a result of the need for two separate vessels, which may go some way to explain its lack of adoption in the commercial sphere.

Wetness

Anaerobic digestion is a fundamentally wet process; MSW-derived biowaste is then, generally too dry to undergo AD without the addition of a suitable liquid. Accordingly, water, recycled post-AD process liquor or some form of liquid biowaste (typically sewage or agricultural slurry) are variously used in commercial plants. One consequence of increasing the wetness is the concomitant increase in digester volume required to treat the material, which has led to a general swing towards relatively 'dry' systems, operating at around 25–40% total solids. However, though there is an obvious financial saving in a reduced-size digester vessel, the thicker contents in such a dry process require more energy to mix effectively, making the running costs higher.

Arguments in favour of the 'wet' or 'dry' approaches tend to end up rather like the ones advanced over mesophilic or thermophilic processes, in that there is no clear overall winner, with each system having particular advantages and applications for certain kinds of biowaste, while a rival technology may be more suitable for others. Thus while it may well be that as a general rule a total solids content of 36% can be looked on as the sensible upper limit for MSW-derived biowaste, there will always be occasions when the specifics of the waste itself or of the given process make either a higher or lower figure the optimum for sustained, stable digestion.

Feedstock Characteristics

The whole point of the biological treatment of waste is to bring about the stabilisation of the biodegradable elements within it. While, as has been discussed in earlier chapters, biowaste is essentially biodegradable, the ease with which this breakdown takes place depends on the specific nature of the particular materials present and their relative abundance in any given sample. This applies not only to less readily putrescible substances like lignin, but also to non-organic waste which may arrive at the digester. Since the inclusion of material which can never be digested tends to interfere with the homogenisation and mixing of the reactor contents, as well as taking up valuable vessel volume to no avail, in commercial operations the feedstock is generally optimised prior to being placed in the digester.

Typically, such optimisation has taken the form of either selecting, or arranging for, an input of source-segregated biowaste, or using some form of pre-treatment on the un-separated (so-called 'commingled') MSW delivered to the plant. Anaerobic digestion is viewed as an appropriate treatment for this latter material far more commonly than composting, for which the purer input tends to be seen as essential. Although there is some merit in this view, the use of hydropulpers and rotary drum pulverisation, with or without the addition of magnetic separation, have made it routinely possible to prepare a suitably sized and consistent feedstock, which is free enough of non-organic inclusions for the digestion process. Devices, like hydropulpers, which make use of water to remove the contraries from the input material are particularly efficient in this role, additionally cleaning the biowaste, as well as simply extracting it. While some companies have trumpeted their claims regarding the achievement of staggeringly high recovery rates and the levels of purity obtained from mixed MSW by totally mechanical means, as yet, no commercial applications nor realistic demonstration plants have been built to prove the case conclusively. Though the place of the truly 'dirty' Materials Reclamation Facility (MRF) will be discussed in a later chapter, in the absence of reliable, representative data on the biowaste fraction from such a device, how well it might perform on a continuous operational basis is open to conjecture. It remains likely, then, for reasons of efficiency and lower investment, that the established methods for pre-treatment will continue to dominate the practical applications of AD for some time to come.

Digester Loading

It is obvious that the actual biowaste loading in the digester is of considerable importance to both that system's efficiency in its own terms, as well as to the effectiveness of AD as a means of material stabilisation. Equally clear is the fact that the loading, in absolute terms, depends on the input waste characteristics, relative wetness and system design parameters, including digester volume and the intended retention period. Often stated in terms of Chemical Oxygen Demand per cubic metre of digester void-space (COD/m^3) or per unit time, in the case of a continuous or semi-continuous process (COD/day, COD/hr), the reduction of organic loading, as digestion progresses, is dependent on the methanogenic stage. Hence, the production of methane is, effectively, the limiting factor, with digesters designed around a high organic loading requiring a large resident bacterial population to achieve proper through-put.

Bacterial Population

The particular micro-organisms present in a digester will depend on the operating temperature range and the design of the system. Clearly, single-stage digestion requires all the relevant groups of bacteria to be present within the bioreactor vessel, while multi-stage systems permit the optimisation of conditions for the needs of the differing microbes involved in the process. The internal mixing of the digester contents, as described previously, assists with the maintenance of the bacterial population, while additionally facilitating their access to the available nutrients. A properly acclimatised culture is, obviously, of some value and consequently, in many commercial systems, the use of process liquor to increase the wetness of new biowaste input also has the desired effect of providing an inoculum of required bacteria. An alternative method employed by some other operators uses animal slurries or sewage to both wet and seed the material, though care must be exercised that this route does not also introduce significant levels of heavy metals into the digester, which may be AD inhibitory and reduce the value of the final product.

Chemical

pH, Alkalinity and Volatile Fatty Acids Concentration

These three factors and their effect on anaerobic digestion are interdependent and, accordingly, are best considered together. Monitoring the pH is required to enable adequate process control in order to provide optimum conditions for the balanced growth of the many kinds of necessary bacteria within a digester. Without this control, although the acidogenic bacteria, having better tolerance for a low pH environment, will continue to produce acids, the methanogens are inhibited and ultimately process failure will result. The inherently variable nature of incoming organic wastes into a digestion system makes exact prediction of pH shifts and any buffering requirements difficult. In some cases, pH control will only be necessary during the start up phase and in overload conditions. However, where values are habitually shown to tend to the low, continuous control may be required. The pH of a liquor undergoing anaerobic treatment is related to several different acid-based chemical equilibria, and it is this which provides the anaerobic system with some capacity to resist pH change. The major chemical mechanism responsible involves the reaction of ammonium ions with bicarbonate ions to form ammonium bicarbonate near pH7 and the interaction of carbon dioxide. For this reason the most appropriate type of monitoring is directly within the reactor, by means of *in-situ* probes, since on exposure to the atmosphere, the digester liquor rapidly loses dissolved carbon dioxide and the pH accordingly rises, making manual extractive samples unreliable.

Additionally, though the ammonium ion is a major source of nitrogen for the methanogens present, when dilution levels are relatively low, as in high-solids digesters, there is an increased potential for ammonia inhibition and consequent salination. Ammonia toxicity is related to pH, since below 7.5 the NH_4^+ ion predominates, which is less toxic than ammonia as NH_3, the form which prevails at higher pH levels.

High alkalinity levels in the range of 1000–5000 mg/l generally provide good safety margins against sudden increases in volatile fatty acid (VFA) concentrations and, in general, levels greater than 500 mg/l are taken as indicative of adequate buffering capacity. It is possible to supplement the alkalinity, thereby readjusting the digester by the addition of lime when the VFAs produced exceed the natural buffering ability of its contents, but this is only effective when the pH is likely to drop below 6.5 and even then, the amounts added should only be sufficient to raise the pH slightly above this value. Above pH 6.8, additional dosing leads to the lime reacting with the carbon dioxide in the biogas produced, forming relatively insoluble calcium carbonate and this precipitates out in the digester without producing any further increase in alkalinity.

There are two main disadvantages of this. Firstly, it removes carbon dioxide which can be an important source of nutrient for methanogenic bacteria and secondly, it increases the likelihood of lime-scaling within the pipework and on the heat exchanger surfaces. There is, moreover, the danger that the removal of carbon dioxide in this way can reduce the overall amount of this gas present to such a point that an increase in pH rapidly takes place, thereby acting against the intended aim and with the potential to disrupt the digestion process.

The concentration of VFAs is one of the most important parameters for monitoring, not least because elevated levels are characteristic of an AD process becoming unstable and they are often, therefore, the first indication that all is not well. Moreover, this can also indicate operational or procedural problems, since insufficient mixing, overloading, temperature fluctuation or reduced bacterial activity can all manifest themselves as an increase in VFAs and a decrease in pH. The progress of digestion within a sealed bioreactor cannot be directly observed. Consequently, for a commercial operation, inevitably anxious to avoid the costly spectre of lengthy down-time for digester evacuation, and mindful of performance criteria and the like, progress indications from the monitoring regime are highly valuable and the role of pH, alkalinity and VFA concentration in the context of monitoring and process control is one of great importance.

Toxicity

A number of substances have toxic effects and this is not confined to purely 'harmful' chemicals whose poisonous nature is widely known and appreciated; even some required bacterial nutrients may become toxic at high concentrations. Thus, the initial input into the digester may itself represent one of the major threats to process stability. Accordingly, toxicity is a more complex matter than might otherwise be supposed and must, therefore, be discussed in terms of toxic levels, rather than simply in terms of toxic materials.

Moreover, the inherent variability of the organic fraction of MSW makes toxicity prediction a much less simple issue also. It is well known, for instance, that certain ions cause inhibition in excess concentration. However, this effect can be counterbalanced by other ions, known as antagonistic, and it may also be exacerbated by the presence of still different ions, known as synergistic ones. Careful monitoring of cation concentrations is required to enable an effective form of cation balance to be maintained in absolute terms, and while this may be possible in a laboratory setting, it is a practical impossibility for a commercial operation, where more readily accessible criteria must be used.

General discussions of biological waste treatment often make much of heavy metals, and this seems particularly common in considerations of anaerobic digestion. Certainly, heavy metals in high concentrations are toxic to the microbial population required for AD, and to the methanogenic component of it, in particular. There is a distinct hierarchy of metal toxicity, with nickel being the most severe, followed by, in decreasing order, copper, lead, chromium, zinc, cadmium and iron[3]. Obviously, the prevalence and likelihood of concentration within any given digester depends entirely on the nature of the biowaste feedstock. The high position of copper within these rankings has, for example, proved to be an operational concern for some AD plants treating agricultural wastes which have a significant input of pig-slurry, since there are relatively high levels of copper in certain proprietary brands of pig food[4].

The issue of toxicity is further complicated by the fact that widely varying levels of given substances have been reported as having inhibitory effects in different studies. This is not entirely surprising, since most research has, by dealing in absolute individual concentrations, tended to ignore the contribution of synergistic and antagonistic interactions between substances, nor has the significant potential influence of bacterial acclimatisation been fully considered[5].

Anaerobic Digesters

In general terms, Anaerobic Digesters themselves can be categorised in terms of their operating criteria.

One categorisation relates to the solid : liquid ratio, which, as mentioned earlier, leads to the classification of a system as either 'wet' or 'dry', though in some respects this is rather confusing, since all AD systems need a measure of wetness to function. The reality is that with a loading of below around 15% total dry solids (TDS) a digester is termed 'wet', above this figure it is called 'dry'. The terms are relative ones and simply relate to the thickness of the digesting slurry contained within. Another useful distinction which was also touched on earlier is the operating temperature range, which is usually either c.35°C (*mesophilic*) or c.55°C (*thermophilic*).

Digesters also fall into one of two groups on the basis of their regime of loading: 'batch' or 'continuous', though some authorities draw a distinction in this latter category between the truly continuous and those which are deemed 'semi-continuous'. As their name suggests, batch systems are filled in one go with all the biowaste to be treated, together with the necessary bacterial inoculum, sealed and then permitted to digest to completion without further interference. The duration of the process depends on feedstock, temperature of operation and the other variables discussed earlier. On completion, the digesters are emptied, typically either leaving around 10%–15% as a seed culture or using liquor from the dewatering process to wet the fresh input biowaste, recharged with a new batch and the cycle repeated. Though they are well suited to the demands of treating relatively large quantities of biowaste, it is often said that the gas production rates of batch systems are very variable and that the production itself is somewhat unsteady. While there is, undoubtedly, some truth in this, operational experience would indicate that this may be due as much to imprecise process control, as much as to the nature of the process itself. This is most particularly true in respect of total system collapse, which can most commonly be traced to the shock of initial over-loading.

Semi-continuous digesters are fed on a more frequent basis, usually once or twice a day, with the simultaneous removal of material which has already finished digesting. While these systems are particularly suited to a regular and steadily arising waste stream, the digester vessel itself is generally larger than a batch one of similar processing capacity, since it plays a part in gas collection. The biogas production for semi-continuous processes is characteristically higher and more regular than for batch systems, which goes some way towards accounting for their popularity in commercial applications. Most of the 'continuous' digesters in large-scale use are, strictly speaking, semi-continuous ones.

In continuous treatment, the input and removal of biowaste to the digester happens in an unbroken cycle, often with an overflow, or pumped return control mechanism, to regulate the level of waste held within the active processing vessel itself. Thus, these systems receive their waste little by little, spread over time, so that digestion takes place uninterrupted, having no natural end point. There are two main limits to the application of such methods to MSW-derived biowaste. Firstly, while, as was discussed earlier in the chapter, mixing of the digester contents is of pivotal importance in all AD systems, continuous techniques exhibit an additional high reliance on pumping, to ensure the maintenance of proper operation. This must be weighed against the energy made available by methanogenesis, to ensure cost-effectiveness if the gate fee for the particular feedstock type is relatively low. Secondly, by virtue of the high internal fluidity required, such systems are principally of use for liquid wastes, or those with a very low solids content.

The relationship of these various methods of classification is shown more clearly in the following diagram.

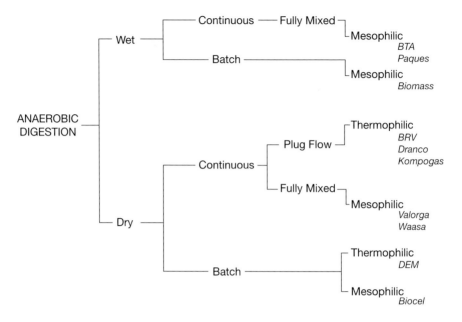

Figure 6.5. Classification of AD by operational criteria

Operational Digester Types

From an operational standpoint, there are several different digester types in use and some of the main kinds of these are listed below.

- Anaerobic Baffled Reactor (ABR) – typically using a horizontal flow of material through the digester vessel, this is appropriate for a wide range of biowaste, and is the basis of the Valorga process, with its patented gas recirculation and mixing system.

- Anaerobic Fixed Film Reactor (AFFR) – limited use for MSW as it is most effective on relatively weak biowastes with low solids content, it works by establishing a microbial biofilm on a fixed growth plate within the digester.

- Completely Mixed Contact Reactor (CMCR) – used for high-strength, and typically industrial, biowaste with the post-processed biomass being recycled after dewatering to increase the solids retention time.

- Continuously Stirred Tank Reactor (CSTR) – used for liquids and slurries, it is effectively the CMCR system without the need for the solids recycle.

- Fluidised or Expanded Bed Reactor (FBR) – this system is only suitable for liquid biowastes, which fluidise the digester's internal bacterial growth medium as they are recycled within it.

- Multi-Phasic Processes (MPP) – these physically separate the stages of AD into different reactor vessels and the main advantages and limitations of this approach were discussed earlier in this chapter, as part of the consideration of retention time. They are very successful as experimental tools for investigations into the nature and pathways of the AD process, but such designs are not generally used in commercial applications.

- Upflow Anaerobic Sludge Blanket (UASB) – with a relatively large population of active micro-organisms, these digesters are best suited to biowastes of low solids content and are, consequently, a common choice for high-strength industrial wastes rather than for use in treating MSW.

These brief descriptions have also served to introduce the final distinction which is sometimes applied to AD systems, namely between those which make use of dispersed-growth bacterial cultures and those which utilise an attached-growth approach to their propagation. Most digesters fall into the dispersed-growth category, with the micro-organisms spread throughout the material undergoing treatment. This inevitably means that an appreciable proportion of the bacteria are lost from the vessel, being swept along with the digestate, either in one instant, at the end of the treatment period, with batch systems, or gradually, throughout processing, in the case of continuous, or semi-continuous, ones. By providing artificial growth media for the formation of an adhering biofilm or establishing conditions as part of the digester design which encourage a settled bacterial blanket to develop, attached-growth systems are less prone to these losses. Accordingly, the mean microbial population density is higher. Moreover, not only does the average residence time for the micro-organisms themselves increase under this regime, but also this longer residence itself largely obviates the typical acclimatisation time-lag often experienced with other systems. As a consequence, attached-growth digesters exhibit greater functional efficiency, since they can be run at higher organic loadings, or with a shorter processing time, their optimum retention period being between 1–6 days, compared with 10–60 days for the dispersed growth equivalent. However, the practical applications of attached-growth systems are almost entirely limited to liquid wastes and, evidently, they are of little use for the higher solid content of MSW-derived biowaste slurries.

Though they clearly differ in certain key aspects, figure 6.6 shows the general features of a 'typical' commercial AD plant.

While it is of obvious interest to a potential user, any direct comparison between the various systems of AD commercially available is, of course, very difficult, since it is possible to advance particular advantages or specific areas of benefit for each. Hence, the manufacturers or operators of a given approach can, truthfully, claim some unique selling-point or distinction over rivals, however limited this may be in absolute terms. Moreover, aside of the variables within each system and the parameters discussed earlier which act as operational constraints on all AD systems, any attempt at comparison of existing plants is further complicated by the effect of different pre-treatment options used to generate the feedstock and the nature of the biowaste feedstock itself. Consequently, data from any given plant can only be viewed as indicative of how a similar one might perform elsewhere, particularly in terms of either biowaste conversion efficiency or biogas generation and quality.

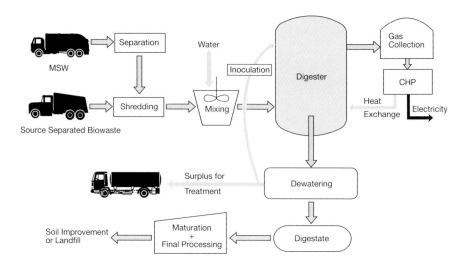

Figure 6.6 Diagram of a generalised AD plant

AD Systems in Use[6]

In 1996, the International Energy Association reported sixty-six digesters in use in the world, of which eleven were pilot plant scale (smallest Stockbridge, UK at 100 tonnes/year, the largest, one of two at Breda, Holland, at 10,000 tonnes/year); one was closed for modification (Dinwiddie, Virginia, USA); and seven of the plants listed as individual facilities occupied only three locations (Balk, Stockbridge and Breda).

Of these, twenty-one were thermophilic, and forty-five mesophilic and the majority ran on source separated biowaste from the MSW stream, either entirely, or using such material in combination with yard waste or manure and/or organic industrial waste.

Only eight took mixed MSW for separation of its biowaste fraction for AD, and, of these, three were only pilot plants and two were reported to be closed. Of the remainder, the waste arriving was not truly un-sorted as we might readily understand it, having first gone through an initial rough separation process in line with French legislation.

The only plant run on mixed waste and sewage sludge was at Vaasa in Finland. An Avecon/Eco-Tech mesophilic digester, utilising the Wabio process, with a throughput of 40,000 tonnes per year, it came on line in 1990. By 1996, this plant appeared to be devoting much of its capacity to deal with sludge.

Of the sixty-six digesters world-wide, over half listed were in just two countries Denmark (18) and Germany (17), with the remainder being in Austria (1), Belgium (2), Finland (1), France (2), India (1), Italy (4), the Netherlands (6), Sweden (4), Switzerland (4), Tahiti (1), the UK (3 – all of which were pilot plants) and the USA (2).

There were a further fifty-one plants recorded in planning in 1996, with twenty more reportedly under construction. Germany and Denmark again made up the largest component, with twenty-two and eight, respectively.

Mass Balance and Product Utilisation

One of the key indicators of success common to any biological waste treatment is its ability to reduce the amount of material requiring to be dealt with by alternative disposal methods, while, ideally, providing a useable product. In the case of anaerobic digestion, this means finding a worthwhile use for both the digestate and the biogas produced. Accordingly, the concept of mass balance is probably more appropriate to AD than any of the other biological processes employed in this role. Indeed, when the digestion plant forms the first point of waste management, receiving unseparated MSW, mass balance considerations additionally provide the primary indication of both generalised recycling potential and of the real, rather than the theoretical, efficiency of the separation system in use. The following diagram and table, based on operational data[7], illustrate the point.

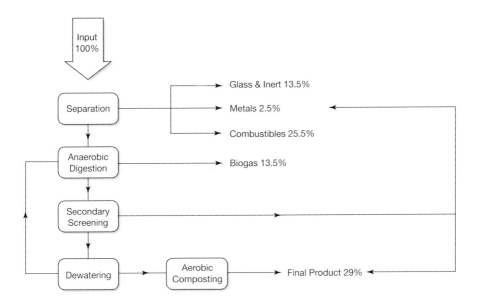

Figure 6.7. AD plant mass balance for MSW

6.3 Table of Original Input Composition

Fraction	Typical percentage
Biowaste	32.5
Glass & Inert	14.5
Metals	5
Paper & Card	32
Plastics	13
Textiles	3

Digestate Utilisation

Irrespective of whichever of the variety of AD systems previously described is used, at the end of the processing, there will be a quantity of residual fibrous material to be used in some way. In commercial applications, this end-use ranges from simple landfill cover, through direct land-spread for agricultural purposes, to the production of a high quality soil additive, typically via a secondary maturation process of some sort. The quality of the digestate is one of the principal issues affecting its use, with public acceptance and economic considerations providing the other main deciding factors. Fairly obviously, the make-up of input biowaste is of paramount importance to the quality of the eventual digestate produced and a clear advantage lies with those plants which take source-separated waste, rather than treating mixed MSW. Though the mass balance shown in the preceding section, for example, indicates a fairly effective segregation, based on the available percentages in the initial incoming waste, the biggest single variable for any facility seeking to take commingled MSW is the efficiency of its separation system. Not all are as good as this one seems to be. Contamination of the biowaste fraction with potentially toxic chemicals and the simple over-representation of non-biodegradable inclusions within the digesting waste mass, both take their toll on the efficacy of the process, which is invariably reflected in the final digestate as the following table[8], for three operations plants, clearly illustrates.

Table 6.4 Digestate Heavy Metals Levels for Different Plants & Waste Types (units parts per million)

Plant Type	Dranco Organic Waste Systems	Kompogas/Buhler	Valorga
Location	Brecht, Belgium	Rumlang, Switzerland	Amiens, France
Waste Type	Source separated MSW	Part source separated MSW	Unsorted MSW
Cadmium	1	0.3	2
Copper	32	40	80
Chromium	23	17	150
Lead	0	40	600
Mercury	0.15	–	2
Nickel	8	15	40
Zinc	180	150	400–750

The situation becomes even clearer if we compare the same processes utilising different regimes to obtain their biowaste input, as table 6.5[9] shows.

The importance of a good feedstock cannot be overstated if the digestate material is to have a real final value. Even though a facility achieves very good overall separation of the incoming waste, as demonstrated by the mass balance shown earlier for the same Valorga plant at Amiens, the fact remains that any biowaste fraction derived by secondary sorting will inevitably have a higher potential for contamination. As has been discussed earlier, the presence, or even the suggestion, of heavy metals within the product seriously limits it end-use. It is worth noting that the AD plant

Table 6.5 *Digestate Heavy Metals Levels for Similar Plants Using Different Waste Types (units parts per million, except where shown)*

Plant Type	Valorga	Valorga
Location	Amiens, France	Tilburg, Netherlands
Waste Type	Unsorted MSW	Source separated MSW
Input tonnage (t/yr)	85,000	52,000
Digestate (t/yr)	44,600	28,100
Cadmium	2	0.5
Copper	80	27
Chromium	150	23
Lead	600	67
Mercury	2	0.1
Nickel	40	7
Zinc	400–750	190

at Borås, Sweden, one of the few two-stage reactors running commercially, takes the purest available source-separated biowaste to ensure the best possible feedstock in the first place, while making use of peculiarities of the split-phase biochemistry to dissolve metals from the digesting mass and leach them out. Thus, digestate quality is improved, with obvious benefits to prospective purchaser confidence[10]. While the two-step approach itself is unlikely to oust the more usual single-stage systems from their commercial dominance, more facilities will have to examine the area of public perception before the major market penetration necessary to absorb the influx of material generated by any upswing in AD technology adoption can be achieved. The need for an accepted common standard for biowaste-derived soil additives has also been addressed previously and though future harmonisation may be a very real possibility, the concomitant issue of application rates must also be considered.

The digestate produced by most operational plants is destined for use as a soil conditioner, which makes the positive contribution of the nutrients present every bit as important to take into account as the negative effects of potential toxic contaminants. Generally, these AD products demonstrate a useable nutrient level, both in terms of absolute concentration and in a form which makes them available for plant uptake. Consequently, in an agricultural situation, the use would result in a reduced requirement for inorganic fertilisers. Moreover, there is increasing evidence that the use of digestates on the land has additional benefits in normal pathogen and parasite suppression, which has previously been reported for composts [11,12].

Biogas Utilisation

In a similar way to digestate, the quality of the biogas produced can affect its final usefulness. It is important to realise that, although biogas from controlled AD processes is similar to landfill gas (LFG), it is quite distinct in quality terms. The removal of the bulk of the inorganic and many potentially polluting inclusions, either by source or mechanical separation, as part of the necessary feedstock preparation to optimise the biowaste input for digestion, also effectively ensures the relative purity of the gas generated. This has obvious implications for its usage, not least since it is common to have to employ high temperature flaring for LFG to destroy the residual pollutant gases, its dirty nature being additionally highly hostile to the fabric of generation equipment. The principal cause for concern in this respect for both AD- and landfill-derived gases is the presence of hydrogen sulphide (H_2S), which occurs as a metabolic by-product of the activities of sulphur-reducing bacteria in the digester. Its abundance in the final gas is largely dictated by the available sulphur-containing compounds in the input biowaste. Gas handling and electrical generation equipment exposed to the acidic effects of H_2S can become rapidly corroded. It is possible to scrub hydrogen sulphide out of the biogas, but the general view is that the use of a lubricant oil with a sufficiently high alkalinity to offer maximum protection, coupled with a regime of frequent oil changes, offers the best commercial cost option[13].

Under AD, given full decomposition and stabilisation of the input biowaste, the maximum theoretical yield is 400 m³ of biogas per tonne of cellulosic material, though around 100 m³ has been shown repeatedly to be a more realistic practical figure, with an energy value typically in the range of 21–28 MJ/m³. In temperate climes, between 20 and 50% of the total energy produced will be required to run the process itself, depending on the specifics of the system in use, leaving the remainder available for use elsewhere.

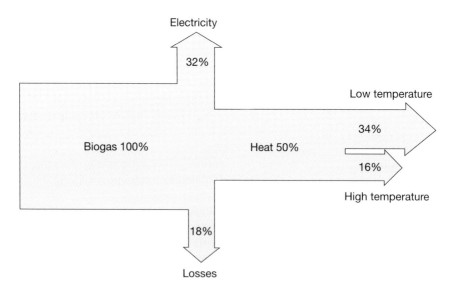

Figure 6.8. Energy flow diagram for gasengine generator set [14,15]

Though it is beyond the scope of the current discussion to examine them in great detail, there are three basic types of engine which are suitable as generating motors for biogas applications, namely turbine, dual fuel and spark ignition and many different manufacturers make examples of each. For any particular AD facility, the quantity and quality of the biogas produced, the nature of the trace gases present, the intended life of the plant, the required noise and air pollution controls and other similar site-specific considerations will largely dictate both the type and the make of engine to be used[16].

Process Liquor: the Forgotten Product

While attention typically tends to focus on the digestate and biogas, as the 'saleable' products of anaerobic digestion, the process liquor which arises is frequently ignored in many assessments of this method as a means of biowaste treatment. From the point of view of both potential client or operator of an AD plant, to do so distorts the full picture and since all commercial waste management needs to be money-making, there are clear implications for operational profitability in hidden costs. Moreover, there are additional environmental considerations associated with the generation of significant quantities of contaminated water, which need to be addressed. This may seem an obvious point, but it is one which has been very publicly missed on at least one occasion, when the early versions of a much-vaunted proposed plant to deal with the biowaste of a certain town in the Home Counties of England had made no provision for dealing with the excess liquid produced. Though the company in question no longer works in biological waste treatment, it would be surprising if they were the only ones ever to fall into this particular trap.

As discussed earlier, in many designs of digester system, some of the process liquor is re-used to wet the newly incoming biowaste feed. Aside of reducing the requirement for, and cost of, clean water, the liquor contains a valuable bacterial population, which acts as a seed inoculum for further processing. Under favourable conditions, this can result in appreciably faster microbial action than experienced during the original start-up. Moreover, acclimatisation of the micro-organisms to their target waste is a significant factor in the efficiency of digestion and there is a body of evidence to suggest that it may also play a major role in toxicity resistance and the maintenance of a biochemically robust digester environment[17]. However, even if all of the liquor so produced were recycled in this way, there would eventually come a time when the concentration of contaminants in the liquid rose to such a level that, rather than acting as a stimulus to optimal digestion, the effect becomes inhibitory. How long it takes to reach this point depends, like so much of the rest of this technology, on operational details of the particular AD system itself, the nature of the waste feedstock, and the efficiency of the separation system, if any, in use. Thus, all AD plants must establish some means for dealing with this spent liquor and any excess routinely produced during normal operation.

Three main routes have been explored by AD operators:

1. Discharge
2. Landfill
3. Beneficial land spread

The first of these, discharge, is often the most convenient option in logistical terms, but it equally commonly comes at a financial cost both for the sheer volume dumped, as well as for any treatment needed to achieve the necessary environmental require ments demanded by the relevant regulatory body. Typically some form of discharge consent will be imposed, setting limits on certain key chemical components of the effluent, dictated by the local circumstances and environmental sensitivity. Discharge to sewer is far more likely to be authorised than to a watercourse, for obvious reasons, though the standards which have to be met even for this can be fairly stringent, particularly where there is little spare treatment capacity available at the sewage works itself. AD process liquors can be of high pollution potential and it is by no means uncommon for them to require a period of treatment to ameliorate them, especially in terms of their BOD, COD, VFA and nitrogen/ammonia levels prior to disposal. Procedures commonly used to do this include aeration, de-nitrification and reverse osmosis, which are well-proven techniques, frequently applied to landfill leachates. However, typical leachate composition, especially for older, established landfills, often differs markedly from AD effluents as the following table illustrates.

Table 6.6 A Key Determinant Comparison of AD Process Liquor and Landfill Leachate All in mg/l except pH (in pH units)

Determinant	AD Liquor[1]	Landfill Leachate
BOD	10,000	1,400[2]
COD	23,000	3,000[2]
Nitrate	50	10[3]
VFAs	5,300	1,500[3]
Sulphate	50	10[3]
pH	5.9	8[2]

[1] Crowland Pilot Plant, average figures
[2] Morris, Steve, *Landfill Leachate Generation & Physical Characteristics*, UKPS Ltd/PCI Membrane Systems Ltd.
[3] Cope, C.B., An Introduction to the Chemistry of Landfill Gas and Leachate, Centre for Environmental Research and Consultancy (CERC)

It must be stressed that the baseline levels will depend on the material and the method of processing, but the following figures from a pilot plant running on various types of largely horticultural biowaste give some indication of the possible input and post-treatment values to be encountered. The aeration treatment used in this case employed a submerged Ultra Fine Diffusion (UFD) system, the effluent being processed in a separate treatment tank, as shown in figure 6.9.

The remaining two options, landfilling and landspread, are far less commonly used. Aside of the similarity of clean-up treatment required to that necessary for the discharge route, one of the main factors against either of these is the transportation cost involved. The same 700-m^3 capacity pilot plant digester, which had run on a typical average content of 15% total dry solids, contained something in the region of 550 m^3 of liquor at the end of processing, the dewatering of the digestate contributing a further 70–80 m^3 to the final volume. In 1997, when this plant closed, the combined treatment and disposal fee was £13/m^3; the transportation cost, using 26-m^3 vacuum tankers, each taking between a half to three-quarters of an hour to fill, was nearly £180 per load, or roughly an additional £6.50/m^3. The final toll involved some twenty-five tanker trips and a total final bill in excess of £12,000, a third of which was purely the haulage charge. This facility was only a development

Figure 6.9. In-tank aeration of post Anaerobic Digestion liquor (reproduced by kind permission of Rob Heap, Bauer UK)

Table 6.7 Horticultural Waste Process Liquor Analysis Before and After 85 Day Aeration Treatment and the Associated Percentage Reductions Achieved[18]. All in mg/l except pH (in pH units) and conductivity (in µS/cm)

Determinant	Baseline	Post Treatment	% Reduction
pH	5.8	8.8	–
Conductivity @20°C	6,950	6,320	9.1
BOD total + ATU	15,800	198	98.7
COD	27,200	1,990	92.7
Solids Particulate 105°C	6,200	28	99.5
Total Dissolved Solids	13,700	293	97.9
Ammoniacal Nitrogen	515	316	38.6
Total Oxidised Nitrogen	1.7	0.3	82.4
Kjeldahl Nitrogen	926	435	53.0
Nitrite	0.79	0.04	94.9
Nitrate	0.9	0.3	66.7
Sulphate	194	63.4	67.3

and demonstration unit, batch digesting a mere 150 tonnes per cycle. Moving the huge quantities of liquor involved for a full-scale, commercial operation would be very expensive and while some plants may move the liquor once to be treated, the prospect of moving it, for a second time, to another site for landfill or landspread becomes commercially limiting. There is a clear advantage to be gained from an on-site treatment capability.

In any case, the dumping of liquid wastes in European landfills will soon largely cease to be an option under the measures to stop co-disposal contained within the Landfill Directive. From 2002 existing sites still taking hazardous waste are required to be classified as such and a total ban will be placed on their accepting liquid waste, though, until 2009, existing non-hazardous sites may offer a possible disposal route.

Land use of waste waters has a long history, though more commonly as a treatment involving controlled applications to effect a form of natural amelioration by biological, chemical and physical processes. One of its major appeals is that it is a less energy intensive method than many conventional treatments. There are a number of regimes employed, including Slow Rate, Rapid Infiltration, Overland Flow and Hybrid processes. These approaches are often applied to sewage sludges, seldom to process liquors from the digestion of other biowaste types. The reasons for this are largely obvious. Digested sludge typically contains smaller percentages of proteins and volatile solids than raw sludge and a higher nutrient level, though the nitrogen, phosphorus and potassium (N,P,K) values are around 20% those of a 'standard' commercial fertiliser. However, this nutrient content is inconsistent and can vary considerably from batch to batch and much of the potential nutrient value may remain in an organic form, requiring additional *in situ* mineralisation before it can be made available for plant uptake.

Another area of concern is the heavy metal content. Certainly for land use, the limiting factor in application rates of liquors with a high metals content often is the cation exchange capacity (CEC) of the soil; hence the annual application rate is typically limited by cadmium content. Where the metals issue is not a particular problem otherwise, the annual application rate is generally limited by nitrogen or phosphoros requirement and, in this case, the total site life by the cadmium loading. Clearly, when the water content of a sludge is purposely low, the hydraulic capacity is very rarely the limiting design parameter. This is precisely the opposite in the case of post-AD effluent.

A number of beneficial uses, especially in forestry, have been successfully explored for sewage[19,20]. Investigation of this avenue for AD process liquors, though of clear attraction as a potential revenue generating 'fertilizer', particularly if chemical consistency could be guaranteed, has been markedly less promising to date. Consequently, at least for the foreseeable future, direct discharge, after any necessary treatment as may be required, seems set to remain the main route.

Suitability for MSW

It is undoubtedly the case that the technology behind AD is able to process MSW-derived biowaste perfectly well and as such, AD meets all the criteria established for an effective biological waste treatment. However, the question of suitability, in the wider sense, encompasses a number of other dimensions beyond what is merely technically feasible. Accordingly, the assessment of the applicability of any biological treatment regime must take many other factors into account, and most of them will be highly specific to the proposed location itself. The amount and nature of the waste produced, local market conditions, existing waste management

Table 6.8 Process Liquor Arrangements at a Selected Sample of Operational AD Plants Taking MSW-Derived Biowaste

Plant	Location	Type	Liquor Arrangements
BTA	Kelheim, Germany	Wet mesophilic	Two-step, on-site denitrification then tankered to local sewage works
BTA	Munich, Germany	Two-stage, wet mesophilic	Two-step, on-site denitrification then discharged to sewer
Dranco	Brecht, Belgium	Dry thermophilic	Aeration followed by discharge to sewer
Paques	Breda, Holland	Two-stage, wet mesophilic	Discharged to sewer
Valorga	Tilburg, Holland	Dry mesophilic	Discharged to sewer

arrangements and political influences may all require consideration as a part of that decision.

While AD has many of these constraints in common with every other form of biowaste-specific process and, indeed, with almost any kind of technology-based waste management at all, there are some which belong to AD alone. One of the single most relevant issues to the adoption of this approach will always be cost, and invariably this will be judged by comparison to the existing established methods and the alternative technologies offered. This inevitably hinges on both current and projected future charges, coupled with the accuracy of predictions regarding the financial implications of medium- and long-term changes to environmental law and taxation, implemented by the relevant legislative body or authority. While certain aspects of this may favour AD, others may not and potential users must take their own view on the likely outcomes and their consequences.

Where the financial model applied to a proposed AD installation differs from that for a composting plant is mainly in respect of the capital and operating costs. Most of the other typical commercial deciding factors, like waste type, minimum tonnages, site acquisition and product market, are essentially similar, though in view of the higher capital expenditure, it is likely that length of contract will often play more of a crucial role also. Since these, and other equally relevant, variables are prone to change over the life of the contract, the position of AD may often be more marginal than that of other, simpler technologies. Nevertheless, where the economic conditions are favourable, particularly when combined with legislative or fiscal support, the particular advantages of anaerobic processing, not least in terms of the enhanced volumetric reduction achievable over other biological treatment methods, can make a very strong case for the utilisation of this approach.

There are limitations to AD, but much depends on what the desired application is intended to achieve. Before an appreciation of the processes involved became more widespread, there was a tendency for AD to be sold as both a method of waste management and a means of significant energy production. Much was made of the maximum theoretical yield of biogas, with the resultant loss of confidence when it became apparent that such figures were not available in a practical sense. It is now universally accepted that systems can only realistically be designed around either waste reduction, or methane generation, and that it is not practical to attempt to optimise conditions for both, within the same regime. This, in turn, has implications for the other major consideration for AD, namely the required quality of the feedstock. If the aim is to maximise the biogas production for conversion to usable heat or electrical energy, the purity of the biowaste input to the digester is not critical, beyond the level required to avoid any possible gross process inhibition. However, for most applications, the main objective is to achieve the greatest possible breakdown of biowaste, and in a manner which produces a useable final product. Consequently, the nature of the feedstock is of much more importance and, as discussed previously, has a direct bearing on its likely end-use. This, of course, raises the issue of source or secondary separation once again, which, alongside cost, is most commonly cited as the cause of reservations regarding the likelihood of anaerobic digestion making any significant increase in its market share.

However, there is a strong feeling that waste management companies, and their clients, who respond rapidly to the regulatory changes regarding biowaste are likely to reap rewards in the years to come and that the exploration and adoption of AD will give commercial advantage in the future.[21]

References

1. Waste Management Paper 27, HMSO.
2. Stevenson, John, personal communication regarding the work on MSW sewage sludge co-digestion done by Thames Water PLC in 1997.
3. Pescod, Professor M. B. (Warren), *Conditions and Variables Influencing the Anaerobic Digestion of Solid Wastes*, Chapter 3 of *Anaerobic Digestion* (The Institute of Wastes Management, 1998).
4. Harper, Dr Stephen R., Project Manager, Engineering-Science Inc., personal comment.
5. Pescod, Professor M. B. (Warren), *Conditions and Variables Influencing the Anaerobic Digestion of Solid Wastes*, Chapter 3 of *Anaerobic Digestion* (The Institute of Wastes Management, 1998).
6. Principal source *Biogas from Municipal Solid Waste* (International Energy Association, 1996).

   The IEA was founded in 1974 as an energy forum for 23 participating industrialised countries. The purpose of the IEA Bioenergy Agreement is stated as '*to increase both program and project co-operation between participants*'. The working definition the IEA places on bioenergy is itself fairly broad, specifically encompassing the production, conversion and use of materials deriving from photosynthesis, either directly or indirectly, for the wide goal of fuel production or as potential substitutes for petrochemical or other energy-intensive products.

7. Delecroix, P., and Saint-Joly, C., *Anaerobic Digestion of MSW – Evaluation of Industrial Facilities After Three Years of Continuous Operation* Proceedings of Biowaste '92, ISWA/DAKOFA publication, 1992, discussing the Valorga Plant at Amiens, France.
8. *Digestate Production and Quality*, Chapter 6 of *Anaerobic Digestion* (The Institute of Wastes Management, 1998).

9. Ibid.
10. Ecke, Holger and Lagerkvist, Anders, *Swedish City Uses Anaerobic Technology*, BioCycle Magazine, June, 1999, p. 56.
11. Keeling, Dr Alan, personal comment on unpublished data.
12. De Ceuster, Tom, J.J., and Hoitink, Harry, A.J., *Using Compost to Control Plant Diseases*, BioCycle Magazine, June, 1999, pp. 61–64.
13. *Biogas Production and Quality*, Chapter 5 of *Anaerobic Digestion* (The Institute of Wastes Management, 1998), p. 29.
14. Nørgaard, P., Vinge, M. and Buchhave, K., *Operational Experience of A 360 Tonne Per Day Joint Biogas Plant*, Proceedings of Biowaste '92, ISWA / DAKOFA publication, 1992.
15. Luning, L., *Aspects of Biogas Utilisation: Practical Possibilities and Limitations*, Proceedings of Biowaste '92, ISWA/DAKOFA publication, 1992.
16. Roughley, D., *Generating Units*, oral presentation at UKPS Ltd/Norweb Generation, Landfill Gas to Power Workshops, (Coventry) 5th October, 1993.
17. Chynoweth, Professor David, personal comment.
18. Heap, R., *Small Scale Aeration Trials of Biomass Post AD Liquor*, unpublished project report.
19. Sopper, W.E., *Maximising Forest Biomass Energy Production by Municipal Wastewater Irrigation*, symposium paper in Energy from Biomass and Wastes IV, 1980, Lake Buena Vista, Florida. Institute of Gas Technology, Chicago Illinois, 1980, pp. 115–133.
20. Bonnin, C., Mailloux-Jaskulké, E. and Luck, F. *Sludge, Resource Or Waste Product: A Look Into the Future*, Proceedings of the Seminar on Sustainable Utilisation of Wastewater Sludge, part of AquaTech 94, ISWA publication, 1994, pp. 127–128.
21. *A Way with Waste. Draft Strategy for England and Wales* (Part Two), DETR 1999, p. 42.

CHAPTER 7
Alternative Biotechnologies

Composting and anaerobic digestion (AD) represent the two general technologies which account for the majority of separate biological treatment of solid waste. Each has a well-established track record and both tend to dominate the wider perception of how best to deal with the biowaste question. However, though both have clear benefits to offer any local authority or other concern charged with the rational management of this kind of refuse material, equally each has its own peculiar set of limitations and requirements.

As has been discussed, under the locally oxygen-deficient conditions of anaerobic digestion, the organic wastes produce quantities of biogas, as a consequence of the biochemical changes within the decomposing material. This biogas has a composition which usually consists largely of methane and carbon dioxide (typically approximately 60% CH_4; 40% CO_2). Though the use of methane to power electrical generators and even its simple flaring, as at landfill sites, are both well known, for certain waste management scenarios, methane production is viewed as a necessary inconvenience in the running of the site. The potential hazards posed by possible methane migration to other areas and explosive gas build up on site are widely recognised, and the estimated costs of measures demanded by some regulatory bodies to mitigate these perceived risks have caused certain otherwise excellent schemes to be abandoned. Also, as with landfill leachate, the production of significant quantities of spent process liquor and the consequent requirement for storage and treatment, with the additional potential for the accidental pollution of ground water, represents another area of concern for both the operators of AD sites and the authorities responsible for their licensing. Much depends on local circumstances, but it is fair to say that a significant part of the relatively high capital costs of AD installations result from these factors and clearly, this has an influence on both the charges to the user and the overall commercial viability and competitiveness of the project.

Nevertheless, the principle of providing idealised artificial conditions to bring about the managed decomposition of wastes and the regulated evolution of biogas remains a well-established technique. As the previous chapter showed, a number of types of anaerobic digester exist which promote this controlled break down in various ways and the operational worth of the AD process has been successfully demonstrated over many years. The volume of material which emerges from such a regime of treatment is typically much reduced compared with the input biowaste and this makes AD a very attractive option in the light of current moves in respect of the amount of biowaste likely to be permitted to enter landfill in the near future.

Despite this, AD suffers from the commercial limitations of cost, imposed by their need for adequate gas seals to prevent air ingress, gas handling to effect the safe management of the potentially explosive methane evolved and internal environmental control, since the majority of the anaerobes involved in the process are relatively intolerant of changes in their surroundings. Moreover, digesters have certain inherent limits on what constitutes an acceptable level of biowaste contamination from other waste fractions. This may be of particular relevance to the application of AD technology to mixed MSW, especially since the standard of separation available from the current best of the 'dirty' MRFs has recently been openly described as 'questionable'[1].

Composting is an exothermic, aerobic decomposition process, requiring the presence of oxygen, the effectiveness of which has been proven over centuries of use. Though the biowaste volumes involved in municipal applications are obviously far larger than in the traditional gardener's heap, the main problem in the scale-up, namely the provision of an adequate oxygen supply, has been successfully addressed in commercial operations by a variety of means, as described earlier. There is an extra cost element as a result of both the energy required to effect this, as well as with the higher materials handling requirement, but these considerations do not appear limiting, particularly for the 'simpler' forms of composting typically used at local authority sites.

However, the major practical limitation on composting as a sole means of mass biowaste treatment is that it achieves less of a reduction in the biowaste volume than is routinely delivered by anaerobic digestion. This is not a problem in the context of an individual garden, where the product is likely to be destined for direct use. However, for a municipal scale composting operation, where the volume of biowaste initially requiring to be processed is very much greater, and the consequent amount of compost produced correspondingly large, the marketing or disposal arrangement for such amounts of derived material may be less easily managed. Moreover, the relatively lengthy retention period for material undergoing windrow or static pile composting inevitably necessitates a large land bank for the operation. Though, as has been previously discussed, it is possible to reduce the processing time by use of more intensively engineered composting systems, these gains are offset by the higher costs in installing and running such plants.

Accordingly, these natural drawbacks have led some to examine the potential of other approaches to biological waste treatment, based on alternative technologies. It is, at the same time, both a strength and a weakness of biowaste management that local conditions can so comprehensively alter many of the factors which dictate the acceptability or otherwise of a given method. A strength in that it encourages the development of a wider base of techniques, providing, to extend an evolutionary analogy, a set of specific niches for novel, or lesser-known, technologies to exploit, and the right set of circumstances for them to thrive, albeit in a modest way. A weakness, because it can deny those who must make the decision as to which approach they will use, for the generally lengthy period of the contract's life, a simple point of comparison between rival technologies. The 'state by state' nature of the chosen response to biowaste diversion, evident in both the US and the EU, is, perhaps, the clearest illustration of this, the goal is set, but the individual mechanics of achieving it is expressly permitted to retain the vital element of local influence. It is difficult to see how this could be otherwise.

Annelidic Conversion (AC)

One alternative which has been explored by a number of companies and researchers across the world is the use of annelid worms to breakdown biowaste, a method widely known as *worm composting* or *vermicomposting*. Using a variety of worm species, the waste is generally treated in a manner broadly reminiscent of composting, hence these commonly used names. However, this itself has been the subject of some debate, with rival terms such as 'vermiculture' or 'annelidic conversion'[2] being put forward as better descriptions of the process. There is some merit to these suggestions, since, although the 'composting' label has a familiar and friendly ring to it, the details of the process, both biologically and operationally are quite distinct from the production of 'true' compost. For one thing, the biowaste is laid in much shallower layers than in a windrow or static pile, often being deposited on top of a soil or true compost bed. This is an important difference and clearly avoids any tendency for the whole mass to self-heat. While worms can be encountered even in thermophilic piles, they avoid the genuinely thermophilic core, being found at the significantly cooler edges of the heap. Though they do require a degree of warmth to remain active, typically above $10°C$, worms do not tolerate temperatures above $30°C$ particularly well, and are killed above $35°C$. A range of between $18°C$ and $25°C$ is generally regarded as being the optimum for most species. Clearly, the highly exothermic conditions of a compost pile would be far beyond their ability to survive. Also, while a healthy community of micro-organisms in the worm beds may play a role in the breakdown of the waste material, their role is an ancillary one, with the main decomposition being brought about by the direct activities of the resident worms themselves. Although worms of one form or another may be present in traditional compost heaps, the numbers are many magnitudes smaller than the artificially high populations deliberately established and encouraged in AC beds.

One of the major advantages of annelidic conversion is that, like composting, it can be scaled to meet particular needs and has been promoted in various forms for both domestic and municipal applications. However, particularly in the case of home bins, this has not been without some problems, since all of the potential pitfalls regarding system design, householder compliance and diligence apply even more rigorously to this approach than to traditional composting. Accordingly, the drop-off rate in use has been reported as 'considerable' by a number of recycling officers, though others have found these bins widely welcomed and well used[3]. For larger amounts of biowaste, worm systems can be tailored to suit, though as a result of the constraining need for the beds to be shallower than windrows, in order to accommodate an equal amount of material, the land requirement is much larger. With a bed area of around $45m^2$ demanded for each tonne of biowaste deposited per week, a typical annual load of 4000 tonnes of garden waste, allowing for the seasonal nature of its production, would require around half a hectare, or one and a quarter acres, of ground simply to provide the beds themselves. To allow for service access between and around the wormeries, more than doubles this figure as the overall total requirement.

With an initial population density, at set up, of five hundred or more animals per square metre and a weekly cumulative production rate of around $0.07kg/m^2$, once established, vermiculture is, clearly, a biomass-intensive treatment technique.

This means that the optimisation of the worm conversion process necessitates careful control of the environmental conditions within the beds, since the physical and biological needs of the worms responsible for the decomposition lie within more precisely defined limits than those of the microbes of composting. As mentioned earlier, the acceptable temperature range imposes its own constraints on bed design, while the heavier demand for oxygen additionally favours the establishment of a large surface area to volume ratio, allowing aeration of the surface layers, where many of the species of worm used preferentially reside, to proceed more readily. While adequate moisture is essential to permit the gaseous exchange function of the worms' skins to be properly discharged, waterlogging is a double hazard, since it both reduces aeration and may actively drive the worms to abandon the treatment bed in search of drier surroundings. In order to meet these environmental conditions, many operators of large-scale facilities employ extensive drainage arrangements and well-ventilated covers. The latter has the added advantage of producing locally darker and less weather-disturbed conditions, particularly in outdoor settings, which encourage the worms to be active for more of the day than might otherwise be the case. This increased activity itself has the added bonus of augmenting the aeration of the bed, as the animals burrow through the deposited material and the underlying layers. In turn, the worms' tunnelling has been shown to have a beneficial effect on the minimisation of odour, reportedly reducing sulphide concentrations in the decaying biowaste by as much as a factor of a hundred[4]. In an effort to maximise the potential for process control, while simultaneously deriving the benefits of a modular and highly portable system, some organisations have managed to developed in-vessel worm bioreactors. This largely removes the need for a permanent installation, which may be a deciding factor for some applications, though it is achieved at a consequently higher unit cost than for the simpler land-based method.

There are a number of species of annelid worms used for vermiculture, which can be divided into two broad categories, earthworms and redworms, though there is some uncertainty regarding the absolute validity of this distinction. Earthworms are burrowers and generally do not assimilate the biowaste organics directly, tending to consume more dead biological material from the soil and passing on the nutrient value in soluble form in their castings. Redworms, also sometimes called 'manure' or 'compost' worms, by contrast, assimilate the biowaste more directly and rapidly, eating half their own body weight, or more, per day and turning the derived nutrients into worm biomass, both their individual growth and overall numbers increasing accordingly. Different species of earthworm have been used with varying degrees of success, including the 'type' genus, *Lumbricus*, *Amynthas* spp. and the deep-burrowing worm, *Pheretima elongata*, native to India, which played an essential part in a Bombay plague prevention project. Set up in 1994, after an earlier outbreak of the disease, this was designed to reduce the growing amount of waste in the city, which had been strongly implicated in attracting and harbouring the vector rats[5]. However, in general terms, annelidic conversion tends to rely on the redworms, with species of *Dendrobaena*, *Helodrilus* and especially *Eisenia* featuring heavily in such operations. These worms are natural inhabitants of the surface layer of fallen plant material and are commonly associated with the production of leaf litter in forest and woodland habitats. Their use in the artificially constructed conditions of vermiculture, is, then, little removed from their role in nature. Consequently, when the bed environment is properly managed, redworms are remarkably efficient at decomposing and mineralising the nitrogen content of biowaste.

The idea of employing worms to biodegrade waste is not a new one, though it does seem to be prone to periodic revivals to fashion. There are three main advantages to vermiculture. Firstly, the efficacy of the process is high, achieving a volumetric reduction often of 70% or more and producing a well-stabilised product. Secondly, it has been estimated that for every one tonne of biowaste laid on the bed, a half tonne of worm casts is produced[6]. Since these are rich in potassium, nitrogen, phosphorus and other minerals, in an ideal form for plant uptake, they are of high fertiliser value. The market for this product has already been established in some parts of the world. Thirdly, the animals themselves offer a potentially harvestable biomass resource, either as seed cultures for other vermiculture operations, or for the fishing bait market. However, this latter outlet has tended to be viewed with some scepticism after the collapse of a Californian pyramid franchise scheme in the 70s and similar more recent bad publicity surrounding certain operations in the UK. However, in the US, Britain and elsewhere, there are a number of long-established businesses which trade in live worms for various purposes, which would seem to indicate that it is possible to turn their production into some form of contributory revenue stream.

Recent work on *Lumbricus terrestris*, a true earthworm, seems to suggest that there may be another benefit to the whole business of vermicultural biowaste treatment. It has been known for many years that this species makes direct use of microbes encountered as food, with a decrease in certain strains of bacilli passing through its gut having been established as long ago as 1950.[7] Though some have wondered if this might help reduce pathogens during processing, generally the lack of the sustained high temperatures of true composting poses potential difficulties for annelidic conversion in this respect. However, in respect of the other major product quality issue of metals and other contaminants, it would appear that vermicomposting may have the advantage, since on some contaminated land sites, *L. terrestris* has been shown to have amassed significant levels of both metals and organic compounds, in particular polycyclic aromatic hydrocarbons (PAHs).[8] This would seem to imply that there is the potential for such contaminants to be bioaccumulated in this species of worm. If, as some researchers suspect, this reflects a wider tendency for annelids in general, the advantage to biowaste-derived 'compost' production is clear, though this might be less welcome news for the live worm market.

The use of annelidic conversion in sequence with other biowaste treatments is an area which appears to have some potential and may have increasing relevance to the wider needs of biological waste treatment over the coming years. Following a shortened period of traditional composting with a final vermiculture treatment has been suggested as one way of obtaining a quality product more quickly. This approach is not without its merits, particularly if the integration of the two technologies can be managed such that significant overall synergy can be achieved, by utilising the respective strengths of each to the full. Annelidic conversion has been shown to be an effective treatment method for the stabilisation of a wide variety of biowaste types, including animal manures and sewage sludge, though its principal feedstock in the 'treatment train' scenarios typically envisaged would be food and garden wastes. It is a particularly efficient means of decomposition when these are fresh, but the need for the material to be kept below those temperatures at which pathogen inactivation occurs makes the eventual product sanitisation more difficult to achieve. Combining vermiculture and composting would seem to bring two major advantages. Firstly, the use of a worm treatment stage appears to enhance both the stabilisation and the derived product quality, compared with composting alone. Not only is a stabilised state arrived at more swiftly, but also there is clear evidence that the volatile organic content is significantly reduced when biowaste is treated in this two-tier manner. Moreover, the apparent innate ability of worms to accumulate a number of hazardous substances within themselves offers the potential for a product preferentially stripped of these chemicals in a way composting cannot rival. Secondly, when the vermicultured feedstock is first subjected to a period of pre-composting, the thermophilic phase can obviously be allowed to run its course without fear of harm to the worms, thereby permitting the input material to benefit from established compost sanitisation practice, such as it is. Once this stage has been successfully completed to the required standards, the biowaste can then be taken on to the annelidic phase of its treatment. Pre-treatment has the additional bonus of reducing the worms' ammonia exposure, a chemical to which they are very sensitive. Though this process enables the product to attain a level of sanitisation which vermiculture on its own might not be able to manage, there is reason to suggest that pre-composting does have a negative effect on worm growth and reproduction[9], which represents a reduction in the rate of worm biomass increase. This, clearly, has a slowing effect on the speed of the overall stabilisation of the biowaste and, in extreme cases, this could have a significant impact on the progress of the whole process. It has been shown experimentally that the enhanced levels of waste stabilisation achieved in worm-based waste treatment are only characteristic of conditions in which a high resident worm biomass is attained[10]. While an obvious and beneficial mutual compatibility exists between vermiculture and traditional composting, it would seem that the primary compost period should be no longer than the minimum required to achieve proper pathogen control of the input biowaste.

For facilities where such a integrated treatment train approach is desired, the use of some form of in-vessel system might prove of great value, since the land requirement of a combined windrow and simple worm facility would typically be enormous, though the additional cost of the bioreactor would equally have to be taken into consideration. As an alternative, some operations have attempted to reduce the ground required without adding to the expense by utilising a standard windrow design, to which worms are added after the thermophilic stage has finished. These have met with varying degrees of success, though the rough handling meted out to the worms as the piles are mechanically aerated, often using the shovel of a front-end loader, has been found to represent a potentially major detrimental effect.[11] The system synergy to be obtained from an optimised combination of these technologies in general would suggest that this approach, though currently little more than an interesting possibility, could assume greater importance. This seems particularly likely if the product quality can be shown to be consistently superior to what is available by other biological treatment routes and though it may appeal to a more limited market, specialist products typically command premium prices. Whether this would offset the additional cost in production sufficiently well to drive a wider adoption of the method, however, remains to be seen.

Ethanol Production

The ability of certain micro-organisms to break down sugars to produce alcohols, principally ethanol, C_2H_5OH, is well known and widely used throughout the world for the production of alcoholic drinks. Fermentative yeasts typically used in this process are poisoned by the accumulation of ethanol to levels above 10%, which means that, traditionally, to obtain higher concentrations, distillation or fractionation is employed. Generally such a process derives a constant boiling point mixture consisting of 95% ethanol, 5% water, though anhydrous ethanol can be produced commercially by azeotropic co-distillation of the hydrated form with benzene or other solvents. Both forms of ethanol make good fuels with excellent combustion properties.

Cellulose accounts for around 50% of the total dry matter in plant origin biomass. A polymer structure of linked glucose molecules, cellulose is potentially a huge, renewable energy store and vast amounts of this material are routinely thrown away, since it makes up the majority of the biowaste component of MSW. However, until recently, the prospect of realising this potential fuel source was viewed as difficult and expensive, the combination of cellulose's resistant links and its close association with lignin discouraging its large-scale hydrolysis to sugars. Early industrial prototype processes made use of the natural enzymes of wood rotting fungi and a feedstock of pulp or old newspapers, though the energy involved in rendering this material into an acceptable form was often a major limiting factor. Taking this one stage further, in the mid-90s, various researchers began experimenting with the idea of genetically modifying waste-indigenous bacteria with the genes of a variety of wood-rotting organisms. Since then, a number of technologies have emerged, based on both whole-organism and isolated-enzyme techniques, which appear to be poised to make the commercial processing of cellulose to alcohol a reality. It is widely expected that the first plant to recover MSW biowaste for the production of ethanol will begin operating in 2001, in Middleton, New York.

There are four stages in the process, called by the trade mark name *CES OxyNol*, which was developed by the Masada Resource Group, Tennessee Valley Authority and Mississippi State University.

Stage 1   *Acid hydrolysis*
Sulphuric acid is used to break down the cellulose into a slurry of sugar water and acid, the solid lignin particles being separately recovered.

Stage 2   *Acid recovery*
The resultant sugary liquid from the previous stage is separated from the process acid, which is recovered and reused.

Stage 3   *Fermentation*
The derived sugar is fermented by yeast into alcohol.

Stage 4   *Distillation*
Which leads to the production of market-grade ethanol.

There is a growing interest, particularly in the United States, in the potential for developing a biowaste-based ethanol industry and a number of individual states have begun to examine their own situations in respect of this. A recent California Energy Commission Report, for instance, records the annual generation of 51 million dry tonnes of biowaste state-wide, comprising forestry residue, MSW and agricultural waste, with a resulting estimated maximum ethanol yield of over 3 billion gallons (US). There are thriving ethanol production plants elsewhere in the US and the world, chiefly deriving their alcohol from primary crop plants, including corn in the Midwest of America and sugarcane in Brazil, where partial ethanol substitution for petrol has been regular feature of life since the 1970s. These established industries can provide useful insight into the practical side of any attempt to extend the large-scale fermentation of biowaste. While there is an obvious appeal to the idea of obtaining readily available, renewable energy from waste biomass in this way, the distillation process necessary to provide this final fuel, itself produces potentially polluting by-products.

For every litre of distilled ethanol made, between six to sixteen litres of stillage are produced, with a characteristically high BOD and COD. Although various uses have been explored for this, including for fertiliser and biogas production, with varying degrees of success, the overall expense of dealing with stillage has typically been high. Recently, the application of digestion technology to this problem has begun to look a more favourable route, principally since the preliminary investigations indicate that this may ultimately lead to significantly reduce costs. In the future, it may be that ethanol production from biowaste will form the first stage of a treatment train, possibly involving the production of biogas from the stillage, with some form of final aerobic amelioration, in an integrated process, based at a single site.

Liquid fuels are of great importance because of the ease with which they can be transported and handled, and their controllable combustion in engines. Ethanol may be either a direct replacement for petrol, or an additive in it, and though its calorific value is lower (24 GJ/m^3, compared with 39 GJ/m^3 for petrol) its combustion properties are better, which largely offsets this discrepancy in practice. Local conditions greatly affect the production costs of ethanol and the market price of alternative fuels, with government policy and taxation instruments playing vital roles in the viability of any commercial ethanol venture. In Brazil, although the use of alcohol/petrol blends dates back to the 1930s, the real motive force behind the wider acceptance of 'gasohol' was the energy crisis of the 70s. In what may now be viewed as a fortuitous combination of events, rising oil prices occurred just as the sugarcane industry, which had made major investments in modernisation, faced collapse in the wake of falling sugar prices. The production of fuel from the crop biomass became a logical step. In many respects, a similar situation exists today with biowaste, in that an abundant and readily available potential source of fermentable material exists and the finite nature of petroleum is well appreciated. The earlier technical difficulties in cellulose fermentation would appear to have been overcome and the first plant to extend this technology to the organic component of MSW will shortly begin operation. Whether the wholesale fermentation of biowaste subsequently takes off will largely depend on the outcome of the Middleton project, and others like it, though it seems likely that wide-scale success will require government backing and the right financial context. Given the fact that this approach could play a major part in addressing two of the largest environmental issues of our time, energy and waste, it seems hard to imagine why it should not be given both.

Eutrophic Fermentation (EF)

The term 'fermentation' is generally understood to have one very specific meaning, namely the anaerobic production of various alcohols, chiefly ethanol, as in the application just described. However, the next discussion applies the fuller definition of the word, that being a chemical change or decomposition brought about in organic substances by living organisms or by extra-cellular enzymes of plant or animal origin.[12] Hence, the experimental, wet, in-vessel, aerobic decomposition process devised and developed by the author to investigate the potential for accelerated biodegradation without methane production was termed Eutrophic Fermentation since it relied on biological breakdown in a nutrient-rich environment.

The original development of the process came out of research into a means of enhanced aeration/remediation of post AD liquor, since the particular effluent batch was proving recalcitrant to more traditional techniques. It is well appreciated that introducing air into some kinds of liquid wastes can effect a regulated reduction of potential pollutant concentration and the principle of providing artificial highly aerobic conditions to bring about the managed attenuation of such liquids has been widely established. This is routinely applied to the effluents of many industries, notably within waste management to landfill leachate. A number of such aeration systems exist, which bring about the introduction of air into the liquid waste in various ways. Though differing in specific detail, the available methods fall broadly into one of two general types, being categorised in terms of their operating criteria. Leaving aside such criteria as rate of oxygen transfer or total dissolved oxygen content, aeration is achieved either by diffused air systems or mechanical aeration. Both normally operate at ambient temperature without any deliberate addition of heating. However, although both approaches have been known and successfully used for some time in various forms they have certain in-built limitations which tend to reduce their efficiency. These limitations, though beyond the scope of the present discussion, principally involve the length of time required to effect treatment, problems of oxygen transfer, liquid stratification and foaming. Attempting to overcome some of these problems led to the development of a laboratory prototype process for effluent treatment which was eventually successfully trialed at pilot plant scale. This principle was subsequently further developed and extended, applying it in modified form to the treatment of solid biodegradable waste material.

Eutrophic fermentation is an aerobic batch process, with a retention period of 30–35 days. Essentially, the biowaste material is rendered into a fine slurry in water, a specially prepared culture of microbes is added and then it is heated, mixed and aerated within a vessel. The author first devised the combined augmentation /enhancement technique used to produce bioacclimatised micro-organisms for the process some ten years prior to the development of EF, for ecological facilitation in created wetland habitats.

Although traditional composting is exothermic, generating its own heat as the process continues, the addition of heat and the maintenance of an elevated temperature is a necessary requirement of EF. In the laboratory experiments, a variety of temperatures were tried in the range 20°C–40°C, the optimum for processing occurring around 35°C. This is unsurprising, biologically speaking, since microbial metabolic mineralisation of their particular substrates is most efficient when the temperature approaches the optimum for their enzyme activity, which occurs at or about 37°C for mesophilic forms. Thus, elevating the temperature of the waste suspension can be seen to accelerate and facilitate the microbial and biochemical processes of decomposition, which is further stimulated by good intra-vessel mixing and the additional availability of oxygen.

Adequate aeration has been produced in all of the trial versions by means of various designs of diffused air systems. Although oxygen solubility decreases as temperature rises, the direct physical transfer of oxygen appears to be only one of the necessary factors influencing the decomposition of bio-waste undergoing eutrophic fermentation, by optimising and enhancing the local conditions. In any case, it empirically seems that the slight reduction in initial dissolved oxygen is more than adequately compensated for by the greater ready availability of oxygen resulting from the additional aeration.

The suspension is stirred mechanically within the bioreactor, which enhances the effect of both the heating and the aeration by ensuring a good mixing of the vessel contents. This is particularly important as it distributes the microbes, their substrates and available nutrients more evenly within the bioreactor, avoiding any potential problems of localised abundance or deficiency. Additionally, it also appears to obviate the problem of stratification and the formation of dead-zones within the bioreactor. This is when certain areas of the vessel containing the waste slurry neither receive sufficient supply of air directly, nor is the slurry drawn away to be taken to a region of the vessel which is sufficiently aerated. In this way, layers of largely untreated material can persist, despite the otherwise apparent proper treatment of the biowaste as a whole. This is of particular relevance when either the vessel used is relatively deep compared to its surface area, which is the most preferable commercial configuration, or the air diffusers are not placed at the very bottom of the vessel. In this latter case, untreated material can accumulate below the aerators themselves and functional thermal inversions can become established, which further exacerbate the problem.

Keeping the body of the liquid circulating also effectively reduces the problem of foaming which is commonly encountered with diffused air systems. The motion of the liquid within the vessel acts against the formation of surface froth, while at the same time increasing the residence or travel time of each individual bubble passing through the liquid, since its path is not directly upwards, but subject to the effect of lateral flow. This both makes EF less prone to problem foaming and increases the effectiveness of oxygen transfer to the suspension, thereby facilitating the overall process of decomposition, which appears to be brought about by an intimately related series of chemical, biochemical and biological processes. These result in the release of nutrients and other by-products into the liquid and, as the breakdown of the organic material proceeds, and the nutrient levels of the liquid rise, functionally eutrophic conditions are established within the vessel. Under normal circumstances, this would lead to rapid oxygen depletion and system collapse. However, by optimising the compromise between oxygen transfer and temperature within the EF bioreactor,

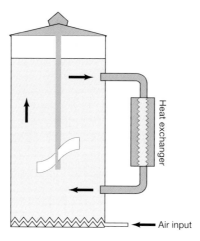

Figure 7.1. Simplified diagram of an EF bioreactor

it has been shown to be possible, at least under laboratory conditions, to maximise breakdown and reduce BOD and COD, while ensuring that the process continues unhindered. This is particularly interesting since, as pointed out in independent review, the overall effect of solids breakdown, which typically results in increased oxygen demand values, would evidently appear to be being counteracted by the rate of assimilation of the available organic matter, indicating 'a highly efficient biological treatment process'[13].

Process Stages

Mechanisms and Pathways

The exact mechanisms and pathways of EF are incompletely understood at this time. What is known indicates that the microbial decomposition of the organic constituents to simple inorganic products, with the concomitant production of energy for self-cell synthesis within EF is broadly defined by the following general equations:

Oxidation/dissimilation
(COHNS) + O_2 + aerobic microbes → CO_2 + NH_3 + other products + energy
(Organic matter)

Synthesis/assimilation
(COHNS) + O_2 + aerobic microbes + energy → $C_5H_7O_2N$
(Organic matter)　　　　　　　　　　　　　　　　　　(New microbe cells)

The Stages of Eutrophic Fermentation

Based on small-scale laboratory investigations, the main milestone stages can be generalised as follows:

Stage 1
Biowaste is slurrified and placed in the bioreactor and the first microbial culture is introduced. Characteristically there follows a rapid biofilm formation and boundary layer foaming, as the pH level falls. This stage is typically completed within 1–3 days.

Stage 2
pH drops to the process low-point and gross biological activity slows or briefly ceases. Dissolved oxygen levels rise and nutrients begin to leach into the liquor.

Stage 3
The pH levels begin to climb during stage 3, reaching a minor plateau at around pH 4.5 which remains from around day 6 to day 21. Functional eutrophication of the liquor continues, but the system is balanced by the abundant availability of oxygen. Dissolved oxygen levels fall and COD, BOD and VFA levels peak.

Stage 4
Biological activity returns as the pH nears neutral and the second microbial inoculum is introduced. The waste material breakdown continues. COD, BOD and VFA levels fall, while the rate of suspended solid formation reaches its peak.

Stage 5
The pH level crosses the neutral zone and continues a gentle rise to plateau at around 8.5 by the last few days of the treatment. Waste material breakdown achieves its maximum operational potential as COD, BOD and VFA levels continue their fall.

Though the whole-process pH range has been shown to be greater than that generally found in AD, EF appears to have none of the typical acid/alkali sensitivity exhibited by the digestion process, which would seem to suggest that it could be more robust in its wider applications.

At the bench (25 litres) and intermediate ($1.5m^3$) scales, the process has routinely achieved a far greater reduction in final solids volume than typical of AD, typically less than 10% of the volume of waste originally placed within the bioreactor and in a shorter time frame. This solid product, after filtration/ dewatering and a period of final stabilisation, might be suitable for land use in some form, though confirming investigation remains to be done. The liquid remaining, which has been characteristically less strong than AD process liquor in the experiments to date, and typically contains between 6 and 10% suspended solids, has been successfully returned to the system, after some further aerobic processing. Alternatively, informal growth trials and independent analysis by the Agricultural Development Advisory Service (ADAS) of the effluent from the intermediate scale reactor have indicated that it has some

Table 7.1 Indicative Key Determinant Analysis

Determinand	Typical End Values	Typical End Range	Whole Process Range	Units
pH	8.1	8.0–8.2	3.5–8.7	pH Units
BOD	1,505	660–2,320	300–15,000	mg/l O
COD	2,035	1500–2,600	1000–11,000	mg/l O
N-NH$_3$	210	150–270	20–600	mg/l N
Nitrate	32	1–35	1–80	mg/l NO$_3$
Nitrite	<1	0.1–1	0.1–1	mg/l N
Sus. Solids	1,350	700–1,700	700–6,500	mg/l
Set. Solids	450	100–600	100–1,000	mg/l
Sulphate	110	70–1,000	50–1,200	mg/l SO$_4$
Sulphide	3	1–10	0.01–10	mg/l S
VFAs	350	150–600	150–3,200	mg/l Acet
Ashed Solids	1,000	100–1,600	150–4,600	mg/l

potential fertiliser use. ADAS assessed samples for key performance indicators such as nitrogen, phosphorus and potassium levels, electrical conductivity, generalised nutrient content and heavy metal residues. Their report considers both the overall fertiliser value of the material, and its potential agricultural benefit, providing definitive recommendations regarding crop-type suitability and likely applications.

With a 'satisfactory' pH of 6, a 'low' BOD of 2,790 mg/l, low ammoniacal nitrogen and the likely slow release of much of the total nitrogen content over several months, ADAS concluded that the liquor was a 'useful source of nitrogen and potash for crop growth'.[14]

Table 7.2 EF Analysis Results (Nutrients)

Principal Nutrients	Units per 1000 gallons*	Comments
Nitrogen (Ammoniacal)	2	Low
Nitrogen	5	Moderate
Magnesium	<1	Low
Phosphorus	1	Low
Potassium	6	Moderate

* Standard UK agricultural measure for fertilizer value

Table 7.3 EF Analysis Results (Metals)

Metal	mg/l	Comments
Cadmium	<0.25	Low
Chromium	<1	Low
Copper	<1	Low
Lead	<1	Low
Mercury	<0.01	Low
Nickel	<1	Low
Zinc	2.7	Low

Viability and Potential for Scale Up

For reasons of availability, it was not possible to run the laboratory bioreactors with actual waste-derived material, so a series of representative substitute feedstocks were made up for the trials from a variety of typical inclusions in the true biowaste fraction. These included vegetables and peelings, plate scrapings/food waste, grass clippings, garden weeds and soft prunings, in ratios similar to those identified from previous 'real waste' analysis. This emulated the biowaste arising from source-segregation or civic amenity sites. To simulate the 'dirty' biowaste fraction produced from mechanical separation of mixed MSW, quantities of newsprint and office papers, plastic (solids and films) and both aluminium and steel beverage cans were added to some of the bench-scale runs, in amounts typically encountered in the field. Although the process was not initially tried on ex-MSW biowaste, it was felt that the close adherence to the character of such material should make the results obtained a fair representation. Moreover, in the deliberately cross-contaminated laboratory-scale investigations, the presence of non-biodegradable material did not affect the processing of the organic matter, simply persisting through, unchanged, to the end, without any noticeable detriment to the biological breakdown. Accordingly, it seems reasonable to suggest that using eutrophic fermentation to treat a poorly separated feedstock would not be likely to lead to significant performance variation, though it would, clearly, have implications for bioreactor volume and design.

Since it was recognised that it is not always possible to reproduce the ideal conditions attainable at bench-scale, intermediate-scale trials were used to confirm the laboratory results, involving 1.5m^3 vessels. These were charged with two kinds of feedstock, namely green-grocer waste consisting of a variety of spoiled vegetables and fruits, and post-producer-sorted biowaste taken from actual MSW. Although the data obtained from these runs indicate that the outcome was not quite as good as in the bench tests, it was nevertheless, fairly impressive and on the whole, the larger scale investigations were independently judged to confirm the effectiveness of the EF process[15].

For aerobic processes like EF, which utilise a completely mixed, suspended growth system, even relatively small-scale studies tend to be reasonably faithful representations of full-scale performance. Since the hydrolysed organics are assimilated by the resident micro-organisms, rather than simply accumulating in the process liquor, the reduced efficiency of the intermediate scale probably reflects less effective reactor mixing and/or aeration. For a scale-up for commercial use, it would be essential that these and other operating features be reproduced at their optimum, which is an engineering matter, rather than one of process biology. By the same token, full-scale bioreactors would need insulation to reduce heat loss, since the energy required to maintain operating temperature would represent a significant element in the running costs of an EF plant.

The viability of eutrophic fermentation as a commercial application has its limitations. Compared with some AD techniques, particularly continuous flow systems, certain process details, chiefly the retention period and the volumes of liquid initially required, have serious implications for the capacity and number of bioreactors required. The power requirement is also heavier than for an AD system of comparable throughput. As mentioned above, heating and heat loss are major considerations, which, together with the energy required to mix and aerate the reactors, and the absence of methane to help offset the cost, contribute heavily to the operating financial equation. The ADAS analysis would seem to suggest that the spent process liquor has a potential value as a source of plant nutrients. It has been pointed out, however, that the availability of nutrients in an unconventional form is no guarantee that a suitable market for the product exists, or could be developed[16]. However, informal discussions with a small sample of farmers has indicated that such a product could have a ready outlet, but only if it were of consistent quality, safe and attractively priced[17]. By analogy to the situation with MSW-derived compost, this should almost certainly exclude the sales of post-EF liquor being viewed as a major commercial profit centre. The fate of the derived solid residue from the process, likewise, would depend on the usual quality criteria.

However, while EF suffers by comparison with AD in some respects, in others it appears far superior. For one thing, no process of anaerobic digestion has shown anything approaching the same level of organic solids reduction, nor produces so relatively pleasant a post-treatment liquor. With so high a level of conversion of organic solids, a stronger final liquid might reasonably be expected, but the BOD was routinely found to be low in the laboratory and this is confirmed by the independent ADAS analysis. The hydrolysed organics from the biowaste feedstock are converted to energy, carbon dioxide and microbial biomass. With a relatively long retention period and the completely mixed, suspended-growth nature of system, endogenous respiration of some of this microbial biomass will undoubtedly occur. Even so, the very low residual solids level at the end of the process is a powerful performance indicator for the micro-organism culture used, strongly suggesting that these organisms are highly effective.

Finally, while the lack of methane does mitigate against recouping any of the operating energy costs, it also means that, if the experimental results could be replicated at full scale, much of the capital expense of intrinsically safe working conditions and highly efficient gas seals could largely be avoided. As an oxygen-rich process, using aerobically respiring micro-organisms to bring about the decomposition of the biowaste, the main waste gas resulting is carbon dioxide. Consequently, without the danger of a build up of explosive methane, an open ventilation system can be used with perfect safety. Openings at the top of the bioreactor would permit the local atmosphere to be actively extracted from immediately above the slurry surface and passed through whatever odour control systems may be required to comply with relevant planning or licensing conditions.

On the basis of the work done to date, eutrophic fermentation would seem to have some merit as a potential alternative to both AD and composting for certain biowaste treatment applications. However, it has not been comprehensively trialed and much of the study involving real MSW-derived biowaste was based on the premise that the feedstock would result from the mechanical separation of mixed waste, by a dirty MRF. Given the current state of this particular art, and the wider view of its 'questionable'[18] present capabilities in respect of achieving sufficient quality of separation, the relevance of this investigation may have been, itself, somewhat lessened. However, there would seem to be a number of issues which would benefit from being further researched, since there are some interesting possibilities for practical applications of this approach, possibly again as part of a treatment train, if the earlier results can be replicated sufficiently well at full-scale. Unfortunately, the company for whom the author developed this process, subsequently decided not to continue extensive research into the biological treatment of waste. Many of the questions surrounding the viability issue remain, therefore, as yet unanswered. This situation seems unlikely to change in the immediate future, since there are unresolved intellectual property matters in respect of EF itself and patent protection has yet to be sought on the original microbial culturing technique.

Changing Circumstances

As has been stressed throughout, it is often specific local factors which can have the greatest effect on biowaste processing and the flexibility afforded by the kinds of alternative biotechnologies described may be vitally important in allowing the need for maximum diversion to dovetail effectively with the demands of a given area. While it is unlikely that either composting or AD will ever establish itself in a monopoly treatment position, it is even more unlikely that any of the currently relatively minor methods outlined in this chapter, will represent a serious challenge to their dominance, on a global scale. However, there would appear to be a place for each of these and it seems certain that the use of such approaches will always have some value in response to particular circumstance. As individual states, authorities and other interested parties begin their considerations of the way they intend to treat biowaste in the future, a role may be found in their plans for these, and other, alternative methods. With adequate political will to achieve the reduction in putrescible waste entering landfill, and suitable financial incentives to do it, it is likely that new technologies will be developed to solve many of the technical problems presently associated with maximising the sustainable re-use of biowaste resources. The question of fuel ethanol production is a case in point. Once the technology is in place, it only remains for the wider economic environment to foster the kind of conditions in which the emergent industry can thrive. The wider issue of resources management is intimately wound up in our approach to waste, a point that will be returned to later. Biowaste management cannot simply be about maximising the diversion of biodegradable material from entering landfill; it must, equally, seek to maximise the rational re-integration of the materials diverted, returning them back into the chain of utility. In this sense, it is no different from any other form of recycling. A recovered aluminium can is of no intrinsic value until it has been made back into another can, or something else of use. There is no virtue in simply having it pulled out of the garbage. This in itself is one of the strongest arguments against those who seek to justify wholesale incineration as an alternative disposal route for biowaste. So-called 'thermal recycling' can only truly be thus described if it does not represent a negative calorific contribution to the overall combustion process. It is apparent that all of the biological treatments have their advantages and characteristic limitations. This may ultimately spawn a series of treatment train approaches, with sequential use made of different technologies to process biowaste, each offering its own contribution step in the overall bio-conversion. This is, perhaps, where these less well-known techniques may best be employed, though they may well retain a sole-use status for certain specialist applications.

In a wider sense, the growth of biotechnology is likely to have major implications for waste management over the coming years, beyond the development of new and better systems of direct biowaste processing. Irrespective of the treatment method used, the clear benefits resulting from a relatively pure stream of biowaste feedstock entering the biological phase of the operation have been well established. This, of course, originally gave rise to the rival views of source separation and on-site sorting as the most appropriate practical way of achieving the required input material. One of the problems with the former approach has been the tendency for the bags in which the biowaste is stored and collected to become a nuisance at municipal facilities, often requiring to be opened and screened out, which can be a labour intensive prospect at this scale. The development of truly biodegradable plastics has already begun to have an impact, particularly at composting plants, where bags which will themselves breakdown, have significantly reduced the amount of work involved. It will be some time before the vision of vast swathes of land growing bio-plastics within transgenic crop plants, with production costs no higher than for potatoes or wheat, becomes a commercial reality, despite its obvious attractions. Nevertheless, progress is being made in this direction with the announcement[19] of the successful use of genetically modified varieties of oilseed rape and cress. Not only is this new material biodegradable, but it is also suitable for a wide number of applications. Although other plastic-growing techniques have been used experimentally in the past, chiefly using strains of bacteria which can produce plastic under certain environmental conditions, the product has proved expensive, costing between three and five times as much as normal oil-derived plastic and typically is too brittle for most uses. By inserting four bacterial genes responsible for plastic production into plants, the expense of feeding bacteria on glucose is avoided, since photosynthesis naturally provides the necessary carbon. At present the yield is low; at 3%, it is around six times lower than has been achieved by other means and while success cannot be guaranteed, the next step will be to attempt to refine the process for greater production.

However, if and when the predicted wider use of a broad variety of new families of such bio-plastics is realised, there may be some unforeseen consequences, particularly in respect of any statutory requirement to reduce the total amount of biodegradable material entering the landfill disposal route. As we have seen, plastics account for something in the region of 8–10% of the waste stream. While it is a thoroughly laudable aim to reduce our exploitation of finite oil reserves for polymer production, particularly when so much of the final product is lost, buried or burnt as unwanted packaging and containers, replacement of even a portion of the total by new forms will inevitably increase the biodegradable component of waste. Dependent on the detail of any legislation regarding biowaste diversion, this may well ultimately mean that, recycling initiatives and waste minimisation aside, an even greater percentage of MSW will require some form of biological treatment in the future.

References

1. *Limiting Landfill: A Consultation Paper on Limiting Landfill to Meet the EC Landfill Directive's Targets for the Landfill of Biodegradable Municipal Waste*, DETR, October, 1999.

2. Worm Digest, No 9, Summer 1995. First attributed to H. Carl Klauck of Newgate Ontario.
3. Informal personal comments to the author.
4. Worm Digest, No 9, Summer 1995, citing work attributed to Astrid Lofs-Holmin.
5. Menon, S., *Worms Recruited to Clear Bombay's Rubbish*, reporting Shantu Shenai's work for the Green Cross Society of Bombay, following an outbreak of plague in India earlier in that year. New Scientist, 17th December, 1994.
6. Denham, C., *Large Scale Profitable Manure Disposal*, Wonder Worms UK, Booklet 2.
7. Day, G.M., *The Influence of Earthworms on Soil Micro-organisms*, Soil Science 69, 1950, pp. 175–184.
8. Hankard, P., Weeks, J. M., Fishwick, S. K., McEvoy, J., Olsen, T. M. E., Wright, J., Finnie, J., Svendsen, C., Maguire, N. J., Ostler, R. and Shore R. F, *Biological Stress Indicators of Contaminated Land; Ecological Assessment of Contaminated Land Using Earthworm Biomarkers*, a collaborative project between the Institute of Terrestrial Ecology and the Environment Agency, Thames Region, (Environment Agency, 1999).
9. Frederickson, J., Butt, K., Morris, R. and Daniel C., *Combining Vermiculture with Traditional Green Waste Composting Systems*, presented at the 5th International Symposium on Earthworm Ecology, Columbus, Ohio, (5–9 July, 1994).
10. Ibid
11. Author's unpublished research.
12. *The Penguin Dictionary of Science*, Fifth Edition, (1979) s.v Fermentation.
    Chambers Science and Technology Dictionary, (1992) s.v. Fermentation.
13. Pescod, Prof. M.B., *Assessment of the Biological Waste Processing Method 'Eutrophic Fermentation' (EF), Developed by Biomass Recycling Limited*, unpublished commissioned report, June 1998.
14. ADAS Environmental (Leeds Waste Group) Analysis Report.
15. Pescod, Prof. M.B., *Assessment of the Biological Waste Processing Method 'Eutrophic Fermentation' (EF), Developed by Biomass Recycling Limited*, unpublished commissioned report, June 1998.
16. Ibid
17. Personal comments to author.
18. *Limiting Landfill: A Consultation Paper on Limiting Landfill to Meet the EC Landfill Directive's Targets for the Landfill of Biodegradable Municipal Waste*, DETR, October, 1999.
19. *Scientists Unveil Plastic Plants* (BBC Online Network) Science and Technology Section, Tuesday 28 September 1999, reporting work by Monsanto, first published in Nature Biotechnology.
    http://news.bbc.co.uk/hi/english/sci/tech/newsid_459000/459126.stm

CHAPTER 8
Thermal Recycling: Energy from Biowaste

The idea of obtaining energy from biomass material has already been touched on in preceding chapters, in the discussions of anaerobic digestion and fermentation. There is nothing particularly new about these methods of utilisation; the production of both methane and ethanol fuels have a long and well-documented history in many regions of the world and under a variety of guises. These are intermediate fuels, derived by biochemical means from the original biological material. However, other approaches, based more on a more strictly thermochemical approach, like direct combustion and pyrolysis, also have their place, and to many, they are the most familiar forms of biofuel. About half of the world's population is reliant on wood or other biomass for domestic purposes, principally cooking, and it has been estimated[1] that the average daily per capita consumption is between 0.5 and 1 kg of such fuel, which equates to around 150 W. This may seem an excessive figure merely for cooking, but it must be remembered that an open fire, the most common method likely to be encountered, has a thermal efficiency of only around 5%.

The energy of all biofuels, both direct and derived, comes originally from incident solar radiation, captured during photosynthesis. The organic carbon material reacts with oxygen during combustion or various metabolic processes to release this energy again, chiefly as heat. The matter of the biomass/biofuel is, of course, simply made available once again for nature's own processes of recycling. Around 250×10^9 tonnes of dry matter per year circulates around the biosphere in one form or another, of which some 100×10^9 is carbon[2] and photosynthesis captures around 2×10^{21} joules of energy per year, or 7×10^{13} watts[3].

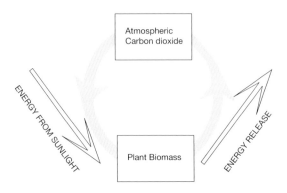

Figure 8.1. Energy and matter in biomass systems

Energy in Biomass Systems

While anaerobic digestion (AD) and ethanol fermentation represent essentially artificial manipulations of fundamentally natural processes, in that the inherent abilities of the micro-organisms involved are used to serve human advantage, the direct utilisation of biomass represents a distinctly different route. Until quite recently, while AD was a technique almost exclusively reserved for biowaste of one form or another, both ethanol production and direct combustion methods were chiefly used on what might be described, in the widest sense, as 'energy-farmed' biomass crops. As has been discussed in the preceding chapter, advances in biotechnology have now extended the potential scope of fermentation beyond expressly cultivated, sugar-rich plants, to MSW-derived biowaste. A similar change may be happening in respect of biowaste as a directly combustible fuel.

Wood and wood-derived materials are the principal components of MSW biowaste likely to be involved in such a move. One of the available options open to a government or authority seeking to minimise the amount of biological-origin waste entering landfill, is to switch this material to incineration. For some kinds of biowaste, chiefly comprising typical food waste, this has often been compared to trying to burn water, since the intra- and extra-cellular moisture content is typically very high in these forms. The natural moisture level present in all biomass fuels may cause significant reduction in the potential thermal output, which has obvious implications for its usefulness. Even purpose-grown energy crops may require a period of drying after harvesting, before they reach their required level of 10–15% water content by mass and the biowaste fraction of MSW is routinely far less easily readied for combus-tion than this. The high energy requirement to evaporate water makes the simple combustion of 'wet' biowaste a highly unfavourable proposition. Incineration is more highly thought of as a means of effective waste management in some countries than in others and for those which do not openly embrace it, any attempt to shoe-horn biowaste diversion into an expansion of the use of mass-burn technology is likely to prove very unpopular. Moreover, within the EU at least, the prevailing driving force of the Landfill Directive expressly sought to avoid such a wholesale swing to incineration-based solutions. This is a point picked up by the UK Government's 1998 Consultation Document, from which it is clear that while an increase in incineration is 'anticipated' the 'Government is of the opinion that incineration with energy recovery should not be undertaken without consideration first being given to the possibility of composting and material recycling'[4]. Moreover, it continues that 'care will be needed to ensure that this expansion does not prevent alternative investment in recycling'[5].

There has been much made of 'combined heat and power' (CHP) installations which gained ground in the UK under successive rounds of this country's non-fossil fuel obligation (NFFO) arrangements, which effectively established a premium rate for electricity produced by renewable means. Nevertheless, there remains a certain resistance in Britain to mass burn incinerators both on amenity and environmental grounds. Increasingly, the wasted opportunity to recycle much of what such facilities combust so profligately has been recognised and incineration without some measure of pre-sorting for recovery is likely to be destined for well-deserved extinction. Despite the CHP hype, only around 7% of waste is dealt with in this way in the UK. Equally, set in the context of their heavy dependence on finite fossil fuels, the huge energy requirements of countries of the developed world demands to be addressed and it is difficult to side-step the issue of the clear resource abundance represented by waste.

The problem is that mass burn incineration, even with the energy recovery which has led to this been described in some quarters as 'thermal recycling', is a disposal route, very much like landfill. Recycling, 'thermal' or otherwise, by definition, is not. Burning 'wet' biowaste at a nett calorific deficit does not constitute recovering and reusing it beneficially; it is a simply an expedient means of elimination. The argument is not one of rival technologies, but rather of diametrically opposed mind-sets and directly conflicting measures of success. The incinerator is paid for the amount of material destroyed; the recycler for the amount saved.

One of the major attractions of both landfill and incineration is that they can each take waste '*as is*' requiring no more of the householder than that the refuse is put out for collection. While disposal dependency formed the prevailing waste management paradigm, the rate of recycling was inevitably destined to remain low, and individual bring systems or separate collection schemes formed its primary route. No matter how much good these do, they cannot make the necessary penetration, in isolation, to make recycling a genuinely viable alternative to straightforward disposal. The balance shifts, however, with a change of emphasis in favour of diversion and the rational treatment of biowaste. As recently as 1996, there was much interest in the development of the 'dirty' MRF as the means to achieve the kind of mass separation of mixed MSW, then thought necessary, to enable serious in-roads to be made into the business of maximising recycling and minimising the amount of biologically active waste entering landfill. In principle, the idea of allowing householders and councils alike to carry on using their traditional mixed-waste collection systems has a logical appeal. The public are not required to change their established habits, neither is the collection authority forced to purchase a new fleet of specialist vehicles, nor re-roster its workforce to accommodate the new arrangements. In practice, however, the quest for a wholly, or largely, mechanical system to sort mixed waste on-site into its component fractions, with anything approaching the necessary purity, has proved largely unfruitful. There have been breakthroughs in some areas, permitting certain materials to be better segregated from the whole, but overall, the dream remains unrealised at present. Perhaps events have overtaken it, and that is how it will stay. As has been discussed in earlier chapters, irrespective of the actual biological treatment technology used, the end results tend to be significantly superior from source-separated feedstock. The widespread acceptance of this, particularly in respect of end-marketability, may ultimately force the acceptance of the dirty MRF approach as one which simply ran itself out of time and missed its chance. The whole clean MRF/dirty MRF thing has been played out before, forming one of those strange cycles of fashion to which the waste management industry seems so peculiarly prone. It would, therefore, be unwise to predict the future demise of this technology too readily, though given its current level of usefulness, it is hard to see how it can make sufficient impact to be assured of continued serious consideration.

With the increased focus on biowaste diversion in the EU, US and elsewhere around the world, some form of separate collection seems a clear front runner to many minds. This also has important implications for the wider sphere of MSW management; it is a relatively small step for householders already separating out biowaste to extend this same approach to any other component of the waste stream which has value, or for which a useful secondary use can be found. If, as seems increasingly likely, the majority of biowaste diversion initiatives necessitate the establishment of source separation arrangements, then it is a simple matter to bring the combustible elements of biowaste under the same regime. At this point, it becomes possible to envisage selective incineration of selected MSW components as an eminently sensible part of a greater, integrated whole. There is nothing wrong with the burning of waste *per se*; the concept of unthinking incineration, however, demands rejection in this scenario. There must be some thought here; some evidence of reasoned and enlightened decision making over the fate of our waste, based on the truly best environmental option, not the merely commercially or politically expedient. The theme of sustainability is one which will be examined in greater detail later, as the single biggest issue in our management of resources – materials, energy and waste – for the millennium. It cannot be acceptable to burn everything in a waste stream any more than it is sensible to attempt to recycle things, like paper, which have little value and for which a cogent environmental argument can be advanced for burning. It is, surely, simply a matter of striking the right balance.

Paper: The Forgotten Biomass

The position of paper in this discussion is a particularly interesting one. While it lasted, the secondary paper price boom of the mid-1990s spawned a widespread interest in paper recycling, and led to the advent of a great many mixed paper collection schemes. With the subsequent collapse of its resale value, the large amount of this material present in the waste stream became something of an embarrassment to recyclers and local authorities alike, since the public had become wedded to the idea of the intrinsic environmental value of recovering and reusing it. To suggest that there was now little point in attempting to recycle paper, which had been the position held by certain quarters of the waste industry all along, was seen as well-nigh heretical. Indeed, for many people, the obligate success of paper re-use was an article of faith and a direct measure of 'green' awareness. Recycling has been described as 'one of the most accessible, tangible symbols of the commitment to do the right thing'[6] and over-zealous environmentalists have been criticised for refusing 'to countenance any argument which undermines their sacred cow'[7].

Paper is often, somewhat conveniently, viewed as wholly distinct from the more obviously biologically active fraction of waste. What must be clearly borne in mind is that the EU Directive expressly defines as 'biodegradable' any 'waste that is capable of undergoing anaerobic or aerobic decomposition, such as food and garden waste, and *paper and paperboard*'[8] (emphasis mine). According to the most recent figures from the Environment Agency, paper accounts for 32%, by dry weight, of the UK's MSW production, which represents its single largest biodegradable component, as defined, forcing the putrescible element into second place by some 11%[9]. Add to this the further contributions from the textile fraction (1%), 'fines' (3.5%), miscellaneous combustibles (4%) and non-combustibles (1%) and the grand total of biodegradable inclusions in the UK waste stream reaches 62.5%, based on the EA figures[10]. Since, on this basis, paper comprises over half of the overall amount, it is clear that no attempt at diverting biodegradable waste can ignore this form of material. With a buoyant market, recovery for resale to the pulp mills is a viable approach; when the more typical commercial conditions apply, segregation, sorting and transport costs can combine to make the whole venture decidedly uneconomic to pursue. Recycling and diversionary strategies drawn up during the former part of the 'boom-and-bust' cycle, are simply untenable in the harsh realities of the latter. For these times, then, there must be an alternative approach, which is sustainable from both environmental and economic points of view.

The generally accepted amount of paper generated per person seems to vary dependent on country and by whom the estimate was made; Europeans appear to produce around 120 kg per capita, while the US figure is nearer double that, based on a rough average of the available data. What is clear is that, firstly, a very large amount of it, especially in volume terms, arises yearly and, secondly, that with a calorific value of around 16 MJ/kg, it compares very well with the kinds of biomass fuels to be derived from short rotation coppicing (SRC). Where paper-biomass scores particularly over willow or poplar is in the speed with which a suitable 'crop' can be harvested. SRC has received considerable favour and interest as a means of providing renewable energy on a local scale. However, such projects are not without their own limitations, not least in ensuring adequate production and continuity of supply. Moreover, the establishment of SRC crops requires both a substantial land bank and a two, three or four year lead-in, dependent on local growing conditions. Such fuels typically have a calorific value of 15 MJ/kg and a yield of between 8 and 20 dry tonnes per hectare per year, or around 3.2–8 tonnes per acre, once established. Apart from some potential habitat/amenity value, SRC is a single-use crop and may have implications for the water balance and irrigation requirements of the area, a point which will be re-examined later.

By contrast, waste paper arises in abundance, and quite routinely, in every community or Local Authority area, thereby offering both a clear continuity of supply, coupled with a production immediacy to the point of use. Apart from the installation of processing and combustion equipment at the user-end, paper 'cropping' for biomass generation has no requirement for additional land use and, obviously, with some 9 million tonnes of waste paper being thrown away each year in the UK alone, no lengthy lead-in period either. Paper has a calorific value of around 16 MJ/kg, as stated earlier, and the yield rate is, on average, 0.35 tonnes per household, or for a typical UK Local Authority annual waste stream of 40,000 tonnes of MSW, around 12,000 tonnes of paper per year and this is available, pro rata, instantly from day one of operation. The biomass to be utilised under this regime represents a dual use crop, firstly as paper *per se*, then secondly, as a derived fuel. There would seem to be general agreement that the basic tenets of SRC are both laudable and sound; surely, then, the idea of growing trees to perform a double service in this way must make even more sense.

An emotive and largely belief-based argument rages over the rights and wrongs of the issue of whether paper is best recycled or burnt, with both sides quoting various statistics, studies and life cycle analyses to support their own view. At least one study showed that when the matter of methane, carbon dioxide and monoxide, sulphur dioxide, oxides of nitrogen and particulates, collectively termed the 'externalities', is taken into account and given a high weighting in the equation, recycling comes off badly. In order for the formula to suggest conventional reuse as the optimum route, these factors must be ascribed very low importance[11]. In other words, this investigation suggests that only if you effectively downplay the environmental costs does paper recycling make sense; burning it is a the better safeguard, if you truly wish to be green[12]. Rival viewpoints aside, there is general agreement that dirty, post-consumer paper is intrinsically unsuitable for recycling. While the fate of clean, separately collected paper may be appropriate to debate, there is little doubt that paper and card from mixed

MSW collections or food packaging from any source, contaminated beyond usefulness to a paper mill, has no real potential for returning into the production chain. Moreover, the balance is further tipped by the nett energy contribution available from paper combustion, which would otherwise derive from fossil fuels and the energy demand and pollution potential of recycling, especially when large amounts are transported around the countryside to paper mills. The utilisation of waste-paper biomass close to the point of its arising and to meet a local need would seem to have much in its favour. It would also close the loop in a particularly apposite way. Most of the Scandinavian pulp mills, from which most of the UK's virgin paper originally comes, burn forestry waste and wood chips to provide their energy, using a locally grown material which, furthermore, is a net zero-contributor of greenhouse gases. Against the right economic and political background, paper too, could equally be a local, carbon-neutral resource; it has been the forgotten biomass for too long.

Refuse-Derived Fuel

The production of waste-derived fuel (WDF) has been tried at many times and in many places, typically being seen, when applied to the total combustible component of mixed MSW, as something of an intermediate between true mass burn incineration and landfill. While, clearly, the paper and card fraction in these fuels would represent a large contribution to the overall burn quality, there is an obvious distinction to be drawn between a 'pure' paper-as-biomass fuel and the significantly more heterogeneous mixture of a typical WDF. Nevertheless, many of the practical considerations are essentially the same and so it is fair to draw certain broad conclusions regarding paper biomass fuels from the greater number of WDF initiatives which have taken place. The physical state of the fuel can either be loose, when it is termed a floc, or densified into pellets. Although both forms may be used as a supplement to traditional solid fuels, or in the case of floc-fuels, fired into specially designed burners, the major problem with their wider uptake seems to be one of marketing. With a calorific value of around half that of typical domestic coal, a significant amount of paper has to be processed out of the MSW mainstream and transformed into fuel to meet the coal-equivalence.

There have been some trials of the pelletisation of paper to provide a suitable fuel. One, at a factory in the North East of England, which produces a well-known brand of tea-bags, was particularly successfully at demonstrating the effectiveness of the general principle, using off-cut bag material. However, this study is not entirely conclusive, since the nature of the paper used in tea-bag manufacture is not truly representative of that likely to be recovered from MSW, being cleaner and having an element of polymer inclusions, though there have been attempts elsewhere to replicate conditions more faithfully. Selection of the pelletising machine to be used appears to be a major factor in the success of the produced fuel. This in itself is not an easy task, since there are many apparently highly similar machines, all of which purport to do the same job, and some of the best currently available were originally developed for the animal feed market, until changes in agriculture across the EU forced them to diversify and develop into a new niche.

The major technology bar, to date, to the development of a waste paper biomass fuel, certainly from mixed MSW, has been the familiar difficulty in obtaining a relatively 'pure' material. For reasons discussed earlier, the utilisation of paper arising from clean, single material collection arrangements, like paper banks or civic amenity sites, has always been possible, but somewhat contentious. To separate out paper from mixed dustbin wastes, though more universally acceptable, is a more difficult proposition.

The waste management industry in general has approached the matter of recovery and separation by developing MRFs. However, the majority of such systems typically cater for dry-recyclables, which have already been subject to a measure of pre-sorting by the householder. The true 'dirty' MRF, which can recover mixed materials from a real, unadulterated, dustbin waste stream, has become something of a Holy Grail for recyclers. Though a number of approaches have been tried, most have relied heavily on human 'pickers' to sort through the waste as it travels along some kind of conveyor system, with all the attendant health and social question marks such an arrangement inevitably raises. The fully mechanical dirty MRF has remained elusive, and, as has been discussed earlier, both in this and previous chapters, it remains a largely unproven technology. Hence, this would appear to be the single largest hurdle to be overcome, before paper biomass fuels can stand any chance of wider recognition, unless, of course, there is a wholesale rejection of the recycling paradigm and the use of clean, pre-sorted paper becomes generally acceptable.

Woody Biowaste

The use of woody biowaste in mulch production has been mentioned in an earlier chapter and in some areas, there is a highly successful market outlet for this kind of product. As an alternative means of use, burning wood is hardly a revolutionary suggestion and there are two main ways in which this form of direct combustion may be relevant to the discussion of biowaste. The first is when the wood to be burnt was itself once discarded as waste, either in a domestic context or resulting from forestry or other wood-based industry. The burning of such commercial wood waste especially has radically revised the disposal economics of a number of industrial operations. The purchase of suitable combustion equipment has been found to amortise over a relatively short period, the combined savings on both disposal costs and the heating bills having made themselves apparent swiftly and significantly. Current developments in this arena, particularly in respect of electricity generation from these relatively small-scale units, seem set to revolutionise the way wood biowaste is viewed still further. However, one of the limitations of this technology relates to handling, which is dependent on the nature of the material itself. Where chips, shavings or sawdust are produced, very effective means exist to convey them into the burner. Rougher wood waste, say from the likes of Civic Amenity Sites, does not easily fit in with this application unless further energy and effort is first put into rendering it into an appropriate form. Consequently, there has been a good deal of work done on trying to find a suitable means of reclaiming the energy locked up in this kind of wood and one of the main technologies widely thought to hold the most promise is pyrolysis.

Pyrolysis

Pyrolysis, sometimes also known as destructive distillation, describes a process in which carbonaceous organic material is heated or partially burnt, in conditions of restricted oxygen availability, to give rise to a secondary fuel and attendant chemical products. It has received particular attention since, although much of the interest relates to possible use on woody wastes, MSW-derived biowaste or crop residue biomass can form the feedstock equally as well. The typical products of pyrolysis include gases, vapours, oils, tars and ash, the exact nature dependent on the particular material treated and the process details. In one form, known as 'Fast Pyrolysis', biomass is rapidly subjected to high temperatures and then allowed to cool, whereupon a dark brown liquid condenses out, which has a calorific value around half that of conventional heating oil and careful management of the process ensures that the yield is maximised. The bio-oil produced is as easy to store or transport as the normal fuel oils it can be used to replace for heating or for use in electrical generators.

Gasification

Gasification is another form of pyrolysis which has received particular attention, since this approach is specifically designed to maximise the production of secondary fuel-gases and, as such, it can be described as a form of incomplete combustion device. This process consists of three stages. Firstly, true pyrolysis which converts the biomass by heat into char and volatile compounds, such as methanol, acetic acids and tars; secondly an exothermic reaction in which some of the carbon present is oxidised to carbon dioxide; thirdly, some of the carbon dioxide, the volatile compounds and the steam produced are reduced to carbon monoxide, hydrogen and methane. Mixed with atmospheric nitrogen and carbon dioxide, the resulting blend is often termed producer gas. The gases produced retain most of the total energy of combustion of the input feedstock, making gasification a very efficient process, typically providing around 80% of the original calorific value in gas form, in the case of wood. Efficiencies of up to 90% can be achieved, for some materials under the right conditions. Though the derived fuel gases need to be cleaned to remove impurities like tars and dust particles, and contain a proportion of carbon monoxide, they are generally more convenient to use, being easier to handle and transport and their combustion can be better controlled. Another bonus is that some of these fuel gases have an appreciably higher energy yield, kilogramme for kilogramme, than the original material, although, obviously, they have a lower total volume. Whereas wood may offer only around 15 MJ per kilogramme, methane, a common product of gasification, can provide 55 MJ/kg. The characteristic calorific value per unit mass of a given fuel is sometimes termed its *energy density* (ED) and a high ED has obvious advantages in terms of storage and delivery.

Having obtained the secondary fuel gases, electrical generation becomes a relatively simple matter, with an overall electricity production efficiency of around 20–25%, and additional heat energy available for other uses. It has been suggested that gasifiers may be particularly appropriate for small-scale power generation, say up to around 150 kW[13], though whether connection to wider distribution grids would be viable at this level remains questionable. However, there is nothing against their use as local embedded power, which is a more rational approach, particularly at this scale, and it is tempting to imagine such units at biowaste treatment plants, providing for most, if not all, of the energy requirements of the site, or even supplying the needs of

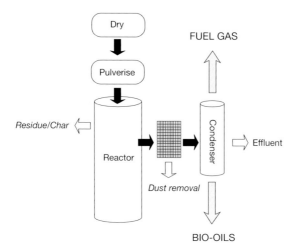

Figure 8.2. A generalised pyrolysis system

adjoining industrial operations. This form of smaller CHP, where electricity and heat are simultaneously produced locally to their users, has much in its favour. It has been estimated that, against the UK norm, it can reduce energy costs by up to 30% and harmful emissions by almost 80%[14].

The production and utilisation of biogas from biowaste under anaerobic digestion has already been discussed in the relevant earlier chapter. Although, as was stated there, the potential for energy production has been sometimes rather oversold in an attempt to make the basic AD technology seem more attractive, it is unquestionably true that this method of biological waste treatment does make a net energy contribution. Even if this is only offset against the running of the plant itself, it still has a bearing on the commercial realities of the operation at the local scale, and implications for fossil fuel reductions on the global one. While an AD facility set up to maximise biowaste reduction throughput will not tend to realise anything like the full theoretical energy potential of the input material, biogas fuel is, nevertheless, an important aspect of the whole system. There is, however, another way in which anaerobic digestion and composting, as the two major biowaste treatments, can play a part in the production of renewable, biomass energy, and that is through short rotation coppicing.

Short Rotation Coppicing

SRC is quite distinct from other, more traditional methods of forestry, and is particularly suitable as an alternative form of crop under intensive arable production regimes, since the plants themselves are often specially bred hybrids, usually of *Salix* or *Populus* species, developed to grow very quickly. In essence, this form of coppicing involves the establishment of plantations and their sustainable harvesting on a regular cycle, to yield useable biomass material on a long-term basis. There are two main approaches to the use of this kind of fuel crop:

- Transportation to a centralised power station
- Smaller scale utilisation on site

There has been some considerable interest in the potential for using the SRC biomass from a number of individual growers in a single mid-range generation plant, but generally the logistics and transport costs of such a relatively low energy density fuel to a centralised facility stand as significant obstacles to its realisation. Consequently, the use of the material more close to its point of production, either to be burnt as part of a site/community heating scheme or gasified for small-scale, local CHP. While it is probable that the loss of energy involved in the gasification process is of similar magnitude to the costs of haulage to a central generator, the environmental costs are far less, since the fossil fuel usage of the road transport element is avoided. When this is coupled with the use of the secondary fuel gases to supply an internal combustion engine as the prime mover for an electrical generator, the environmental savings are even greater, since the engine operates at near-diesel efficiency, but again without the unwanted additional atmospheric contribution of ancient carbon dioxide.

The Scope for Biowaste Product Use

The amount of energy produced in the final crop largely depends on the climate of the growing area, irrigation needs of the trees, the available nutrient status of the soil and the aspects of the husbandry regime. While the first of these, the climate, is obviously relatively fixed for any given geographical location, it is with the optimisation of the three remaining production variables that biowaste may play an important role. As was discussed in Chapter 5 and elsewhere, the use of the eventual product arising from biologically-treated biowaste offers a significant increase in the water-holding capacity of a soil thus amended. As the UK field work reported earlier demonstrated, the addition of around 250 tonnes of compost enables each hectare treated to hold between 1,000 and 2,500 tonnes of rainwater.[15] A considerable amount has been said over the years about the high nett transpiration pull of SRC plantations and the consequent hydrological implications for the land. While there is some reason for concern, it is unfortunate that much of what has made so indelible an impression has rather confused the issue by treating willow and poplar as if they were entirely identical in their water relations, which they are not. Once established, the poplar has a very deep tap root; in situations where grassland might normally support a water table at around 1.5 metres, planting poplar may make it ten times lower. This massive disruption does not seem to happen with willow, with its typically more shallow root system, the water requirement having been estimated as only roughly equivalent to what sugar beet or winter wheat would require[16] under the same circumstances. Nevertheless, even if much of the reputation all forms of SRC have acquired is largely unfounded, a water requirement broadly similar to conventional arable crops still represents a sizeable demand. Obviously, this forms a limiting factor for areas with low water holding capacity soils, which otherwise would be ideal for growing biomass fuels. The evidence from the trials of large-scale, biowaste-derived compost additions conducted in the loose, sandy ground of East Anglia, would seem to suggest that this route would enable SRC crops to be grown without any need of further irrigation in all but the most exceptional of years[17] and even then, the requirement for watering would be severely reduced. These same studies showed that relatively immature composts applied to soils and left to finish off *in situ* are particularly effective in this role, being able to absorb and retain between two and ten times their own weight in water. Similar findings have also been suggested for the dewatered digestate resulting from anaerobic digestion and the filtered solids of eutrophic fermentation[18]. AD digestate sludge can be stabilised after separation from the process slurry by being allowed to undergo a period of maturation under aerobic conditions. This is an approach used by a number of operators and is sometime known as 'secondary composting' though the term is not an entirely accurate one, in the strictest sense. The nett result is a material which retains a desired level of microbiological activity, with a combination of high humus content and excellent water retaining properties, which seem to be at least as good as those achieved by a true biowaste-derived compost at similar application levels. In the small-scale growth trials undertaken by the author, using first-generation clones of *Salix caprea* and *S. viminalis* (Bowles' hybrid), no significant differences were noted between the directly observable growth, dry mass yield or development of willows planted in biowaste-derived, wholly aerobic compost or in aerobically semi-matured digestate. Broadly similar results were obtained using filtrate obtained from the experimental eutrophic fermentation process, which had also been subsequently allowed to stabilise aerobically, though this formed a considerably more limited investigation. Consequently, the results in respect of the post-EF material, though interesting and apparently supportive, remain effectively unverified, as yet. It would seem that biowaste-derived materials such as these, applied in appropriate quantities to soils, either as a surface mulch or a ploughed in initial soil amendment have the ability not only to lower any requirement for supplementary watering of the growing biomass crop significantly, but also reduces any attendant tendency to drought-stress in the plants themselves and reduces soil-nitrate leaching. The ability of coppice crops to reduce the wash-out of nitrates and phosphates has led to their being suggested as potential pollution control measures to avoid agricultural fertiliser residues contaminating waterways.[19] It is not difficult to imagine perimeter belts of willow, performing a similar role on a biowaste treatment site; certainly, the effectiveness of the 'bio-bund', an artificial slope of compost material planted up with willows, to reduce another form of pollution, in this case noise, has been demonstrated. The use of expanses of banked coppicing, grown in biowaste-derived material, both to hold in potential pollutants and cut down on noise nuisance, would seem to be a particularly elegant way to address these two likely planning considerations, for any newly proposed biological treatment plant.

There appears to be some uncertainty regarding how much nutrient is actually removed when the coppice is harvested, with various workers estimates ranging from only around 30 kg to nearer 150 kg per hectare for nitrogen, and other figures suggested of 16 kg for phosphorus, 85 kg for potassium, 12 kg for magnesium and 200 kg for calcium. A UK Forestry Commission investigation into this issue produced nitrogen results of 135 kg per hectare and 16 kg of phosphate. This would represent an annual requirement around 20% that of a cereal crop and it has been consequently suggested that, at this kind of removal rate, nutrients are unlikely to be a limit on growth on fertile sites for at least the first few harvest cycles. However, after this time, or from the outset in soils which have insufficient natural fertility, there will be a supplementary nutrient requirement which could be well met by the addition of suitable biowaste humus material, which, given the added advantage of the water retention described previously, makes a single material application perform a dual function, with clear economic implications for the overall project.

This also has bearing on the last of the three factors listed, the husbandry regime. While it is obviously necessary to fence a newly planted bed securely to preclude damage to the trees as they establish themselves from rabbits, deer and stock animals, one of the greatest causes of poor crop performance, or even outright failure, is direct competition from other plants. It has been estimated that uncontrolled grass or weed growth around the new crop in its first season can reduce its growth by as much as 50% and the dry matter yield by 20% or more. Even once established, a coppice needs some weed control measures, especially after harvest-time, to help ensure continued optimal performance. Generally the preparation of land for SRC takes place in the autumn, which goes someway towards helping control the weeds, but their ability to take hold rapidly come the growing season, and then to compete strongly with the SRC plants for nutrients and water means that this alone is not enough to deal with the problem. It has been widely accepted that one of the best ways to prevent the growth of weeds under coppice conditions is to apply a heavy layer of mulch around the planted trees. For reasons which have, again, been discussed more fully elsewhere in this book, soil-products derived from biowaste are a prime candidate in this role, which clearly supplements and further enhances the benefits in terms of plant–water relations and nutrient availability previously described. Although most of the clones commonly offered for commercial SRC operations have had a measure of disease resistance bred-in, the natural plant pathogen inhibition of biowaste composts may be viewed as a final bonus to their use for this application.

The kind of integrated approach which would link the production of a biowaste-derived soil enhancer to the creation of a biomass-fuel is still in its infancy. Whether on a single site or at separate, but local, facilities, this could represent one of the most effective routes to providing both a powerful stimulus to the concept of unified sustainability in resource management, and a means to establish a very real added-value to the material produced by biological waste treatment.

Biomass and Renewable Energy

Aside of the potential for alternative avenues of biowaste utilisation, the whole discussion of biomass and biofuels centres around the issue of greenhouse gases, global warming and climate change. In the light of the Kyoto Protocol, the fundamental question becomes, should biomass be viewed as an essentially recyclable, sustainable energy substitute for finite fossil fuels, or simply as a long-term carbon-sink? This is no longer simply a matter of interest only for the forestry industry; while inevitably the best environmental use of trees remains of great importance in this context, the ripples now spread wider. It has been suggested, for instance, that landfilling waste paper represents a major route for carbon sequestration, in that, by virtue of its slow breakdown under such conditions, it effectively locks up its constituent carbon for 10–20 or more years. Thus, it is argued, although its combustion would have been carbon-neutral, since it would have released only 'modern' carbon dioxide into the atmosphere, not burning it represents an additional contribution to CO_2 reduction in real-time. This has some appeal, but the balancing view is that when the paper thus buried does eventually decompose, the carbon is released not solely as carbon dioxide, as in straightforward burning, but as a mixture of CO_2 and methane, in which the latter, some thirty or so times more damaging as a greenhouse gas, routinely makes up over half. To take it to its full conclusion, the paper-biomass-as-a-carbon-sink argument only holds together if the landfill gas eventually produced is collected and itself burnt, thereby returning only carbon dioxide to the atmosphere. Simply burying the paper is not enough.

What is true of paper holds true for all forms of biowaste. This, of course, has formed one of the primary underlying drives behind the EU Landfill Directive and is equally fundamental to many other attempts at increased biowaste diversion, around the world. If the idea of carbon lock-up in landfills ultimately leads to an acceptance of the need to burn the final product, then common sense suggests that reclaiming the energy this makes available, is a far more intelligent course of action than simply flaring the methane just to get rid of the problem. However, once we have accepted this, there would seem to be no good reason for landfilling biowaste in the first place, since it is much easier to manage the production of energy under any of the systems described earlier, or others, by means of a controlled process, rather than the necessarily haphazard gas evolution and differential decay rates to be encountered in a typical landfill site.

Moreover, for more than a decade it has been realised that a balanced use of biomass, combining the carbon-sink effect with an on-going replacement of fossil fuels, has obvious advantages over its use solely for carbon sequestration. This is especially true for the likes of short rotation coppicing, which, since it is based on a perennial cycle, may confer significant benefits to the soil, to local biodiversity and in employment, with rural communities gaining jobs tending the coppices, rather than having the land bound up in carbon sinks and effectively unworked.

Our energy requirements are, unquestionably, great. In the United States, where the demand for electricity has grown by an average of 2.7% per year over the past ten years[20], the Executive Order on Biobased Products and Bioenergy, made in August 1999, set the goal of tripling US biomass use by 2010. It has been estimated that this could generate some $15 billion of new income and have the same effect on carbon emissions as removing around 70 million cars from the road[21]. Similarly, a recent European Commission study made the suggestion that the EU as a whole should double the current 6% contribution made by renewable energy sources, also by 2010. It was proposed that biomass energy would provide an additional annual 90 million tonnes of oil equivalent (Mtoe), raising its overall share to 137 Mtoe, of which half of the extra was expected to come from energy farmed crops specifically. It would seem that the door lies open for many other forms of biomass fuels to make up the rest. While it is by no means a solution for all biowaste, there seems to be a very clear opportunity for at least some of the biodegradable material, which future years will see diverted from landfill, both to make a contribution to the world's energy needs and help achieve the targets set at Kyoto. There is a clear sense of aptness about this, since the desires to achieve both the diversion of biowaste from landfill and a reduction in carbon dioxide emissions to atmosphere, are driven by very similar forces and are, essentially, about the same thing. It has been suggested that if we were only able to trap just a fraction of the sun's energy falling on the earth, we could provide pollution-free power for all of mankind's activities across the globe. While we may be some considerable way from making this anything other than a powerful image, photosynthesis remains one of the most efficient ways to capture incident solar energy and, directly or indirectly, waste biomass can have a part to play in this arena.

As with the biowaste conversion processes themselves, there is also clear scope for the amalgamation of approaches and the utilisation of serial, complimentary technologies to derive maximum energy value from the input putrescible feedstock. In this way, wider environmental goals could be achieved than simply a diversion from landfill, with true 'thermal recycling' a realistic additional option in a suite of biowaste treatments, standing alongside nutrient and humus recovery. Hence, it is possible to envisage the concurrent constructive management of biologically active waste material and the production of a significant energy contribution, without automatic recourse to wholesale incineration. For the truly renewable, there must always be a better way than the quick convenience of mere disposal.

References

1. Twidell, J. and Weir, T., *Renewable Energy Resources*, Chapman & Hall, London, 1994, p. 291.
2. Ibid, p. 281.
3. Ibid
4. *Less Waste More Value*, Department of the Environment, Transport and the Regions, June 1998, p. 10.
5. Ibid
6. Pearce F. *Burn Me*, New Scientist 22 November, 1997, p. 31, quoting Frank Ackerman, Professor of Environmental Policy, Tufts University, Medford, Massachusetts.
7. Ibid, quoting Richard Sandbrook, Director of International Institute for Environment and Development, London.
8. *Limiting Landfill: A Consultation Paper on Limiting Landfill to Meet the EC Landfill Directive's Targets for the Landfill of Biodegradable Municipal Waste*, Department of the Environment, Transport and the Regions, October 1999, p. 11, quoting the definition from the EU Landfill Directive.
9. Ibid, p. 12, using figures from the Environment Agency's National Household Waste Analysis Project.
10. Ibid.
11. Pearce F. *Burn Me*, New Scientist 22 November, 1997, p. 33, reporting the work done by Matthew Leach, Energy Policy Analyst at the Centre for Environmental Technology, Imperial College, London.
12. Ibid
13. Twidell, J., and Weir, T., *Renewable Energy Resources*, p. 295.
14. *Cutting Emissions and Costs with CHP*, Green Government, December 1999, p. 20, quoting Sandy Honeyman, Co-Generation Manager for Scottish and Southern Energy.
15. Butterworth, W., *A Top Idea That Holds Water*, Wet News, (Water and Effluent Treatment News), Volume 5 Issue 17, October 1999, p. 4.
16. MacPherson, G., *Home Grown Energy from Short-Rotation Coppice*, Farming Press Books, 1995, p. 175.
17. Butterworth, W., *A Top Idea That Holds Water*.
18. Author's unpublished work.
19. MacPherson, G., *Home Grown Energy from Short-Rotation Coppice*, p. 37 reporting the ideas of Dr Nicholas Haycock of Silsoe College, Cranfield University.
20. Perkowitz, S. *The End of Light As We Know It*, New Scientist, 8th January, 2000, p. 30.
21. Feinbaum, R., *Analyzing the Potential of Biomass Recovery*, BioCycle, November 1999, p. 27.

CHAPTER 9
The Way Ahead

With an increasingly strong legislative force driving biowaste diversion, in the form of the Landfill Directive within Europe, individual state ordinances in the US and similar regulations elsewhere in the world, the clock is effectively ticking for those bodies charged with translating political rhetoric and the law-makers' edicts into practical realities. For those countries which have a traditionally high dependence on landfill as their main waste management arrangement, there will be pressure to take action on two fronts. Firstly, there is the need to limit the use of landfill to ensure that only the permitted quantities or types of biowaste are permitted to enter this route, in line with the relevant legislative targets set, and the appropriate time-frames for their attainment. Secondly, it is essential that landfill-dependent regions begin to seek alternative solutions early enough in the compliance cycle, thereby encouraging the development of suitable biological waste treatment options and/or providing support for minimisation approaches designed to address the very root of the problem, by reducing the amount of biowaste arising in the first place. Certainly, for those EU Member States which fall into this category, the magnitude of the change necessary and the relatively short period available, under the terms of the Landfill Directive, to achieve it, means that very little time can be wasted before beginning the measures necessary for getting these two courses of action underway.

The first of these requires the formulation of a policy which not only enables the goals set in respect of biowaste diversion to be achieved, but also are practical to put into operation, beneficial in its overall environmental impact and both cost-effective and affordable. For reasons that will be developed later, the demands made by the long-term issue of sustainability may also play their part in this process. These requirements can be largely encapsulated by a set of more specific criteria, and any measure intended to address the limiting of landfill must, obviously, be subject to examination under precisely this kind of structure for an objective evaluation of its likely full proposed effect. There are three key areas which need to be considered:

1. Practical aspects, like the certainty of meeting the targets set, enforcement provisions and the direct environmental and economic benefits derived.
2. Administrative issues, including ease of running, the 'transparency' and demonstrable fairness of the measure and the uniformity of its application.
3. Political considerations, covering the wide spectrum from areas of local relevance and accountability, through to wider compliance questions with national and international policy.

It is, equally, clear that any new instrument developed in this context must also fit in with existing views of business and environmental regulation, as well as dovetailing seamlessly with current control measures for planning, land utilisation, permitted development and the management of waste. In this way, there seems little value in significantly altering the existing *status quo* of responsibilities, either in respect of local authority demarkation, or statutory regulatory bodies, since particularly in the case of individual councils or municipalities, the unique advantages they bring in terms of local knowledge makes their contribution to the discussion of enormous value. While it is unlikely that, at least in the EU, expressly local concerns at this scale will dictate the overall approach adopted by individual Member States to meet the Landfill Directive requirements, the co-operation of this level of local/regional government will prove essential in achieving the final outcomes desired.

The diversion targets established by the Directive are set for Member States as a whole, which may be a potential area for some difficulty at least in the early stages of compliance, particularly for countries, like the UK, which has devolved certain major sectors of power to its constituent parts. Thus, the division of these goals will have to be agreed on a national basis and the relevant allocations made between the administrations of England, Wales, Scotland and Northern Ireland. This opens up the very real possibility that different measures will be employed to meet the requirements in the different parts of the UK, which in turn raises a number of issues. One of the main potential problem areas comes as a result of the optional derogation provision within the original Directive. This was discussed earlier, in chapter 3, but to briefly reiterate, any country which landfilled over 80% of its waste in 1995 has the option of postponing the effective date for each of the three targets, by up to four years. This has clear attractions in that it provides a longer lead-in time during which the necessary additional biowaste treatment facilities, collection arrangements and infrastructure can be built up and avoids any compulsion to the wholesale adoption of incineration, as the only other non-landfill technology which could be implemented quickly enough to meet tighter deadlines. While there is a strong argument to be made that the sooner the targets are achieved, the better, there is no getting around the fact that the UK and others do have a heavy landfill dependence and any attempt to meet the Directive targets in the first wave, would almost certainly require a large swing to incineration. For a variety of reasons, this is a generally unpopular method *per se* with a substantial sector of the UK public and clearly not an ideal means of dealing with putrescible waste. Ever since the provisions of the Environmental Protection Act came into force in 1992, there has been a groundswell of interest in biological treatment methods, but despite this, and the certain knowledge of the, then, impending European legislation, no really major in-roads have been made into the wider commercial application of biotechnology to this area. Indeed, there has been some evidence from certain quarters of the abandonment of any further attempts at biowaste research or marketing in favour of traditional recycling, though this may yet prove to have been precipitate and short-sighted. However, it remains likely that even given the additional periods under derogation, biological approaches to waste treatment will only flourish in an appropriate commercial climate.

The Scottish Environment Protection Agency (SEPA) has stated the intention that it will not seek to take up the option to postpone the implementation of the Landfill Directive provisions in respect of biowaste unless it is 'absolutely necessary'.

Clearly, for the situation to be workable, any derogation decision must be taken across all of the UK, since if one of the countries were to opt out, and another did not, significant barriers could be created between the two lands, and this would be most telling on the border areas, where trade traditionally causes an overlap outside of soft 'national' boundaries. This is an issue which may well have implications for some other European countries bordering landfill-dependent neighbours, despite all of the advances in reduced internal frontiers and open trade within the EU.

Much of what needs to be achieved in the early rounds of any large scale, staged reductions in the quantities of biowaste destined for landfill, either within Europe or elsewhere, can be achieved by existing arrangements, often with only a little additional resourcing. It is inevitable that many concurrent factors in modern waste management tend to provide an almost synergistic impetus in largely the same direction. Waste minimisation and recycling initiatives, rising consumer awareness and legislation aimed at reducing packaging, together with the falling availability of void-space and planning pressures surrounding the establishment of new tip sites, all tend to act as natural limits on the use of landfill for biowaste. Coupling this with even modest currently available composting or recycling arrangements and a positive move towards diversional thinking, makes early sector targets relatively easily achievable in most cases. Where the need for the kind of formulated approach outlined previously really becomes apparent is in the meeting of later ,and higher, levels of biowaste redirection, which, of course, also brings us back to the earlier point, that in order to achieve the main goal, it is necessary to settle upon a coherent course of action relatively swiftly and then implement it without delay. The provision of an appropriate framework and the existence of a suitable economic environment satisfies a large part of the requirements for biotechnology-based systems of treatment to be amplified, optimised and developed for commercial exploitation, on the sort of scale necessary. However, there is frequently a considerable time condition in simply finessing new planning applications through to the building stage, in finalising appropriate collection or separation arrangements and establishing a market outlet for the sorts of volumes of final product likely to be produced. As has been mentioned elsewhere, on a number of occasions, each of these is often not immediately straightforward to achieve, providing the opportunity for a further slowing of progress. Even with additional breathing-space built in to the overall scheduling, as in the European derogation option, it may prove difficult for all of these arrangements to be concluded on time. Consequently, for waste management operators seeking to provide appropriate facilities every bit as much as for the governments charged with devising the overall strategies themselves, procrastination may turn out to be a luxury which neither can afford.

Options for Landfill Diversion

There are a number of possible options for limiting the use of landfill for biowaste and the onus of each of these tends to fall either on the waste management industry, and then largely on the site operators themselves, or rest with local authorities. One option which would affect both would be the unconditional banning of all biowaste material from landfill. While this is, clearly, a simple and unequivocal method, it would represent an enormous change at many levels and create major upheaval, particularly for any largely landfill-dependent country. It has some obvious benefits, in that it would place a direct duty on every site, operator and municipality not to allow the burying of putrescible waste, which makes issues of compliance and enforcement, at least, a fairly straightforward matter to administer. It would also bring about a reduction in methane production, though given the inevitable time-lag between biowaste deposition and biogas evolution, the benefits of this would not be felt for a little while. While it would be a relatively simple matter for the due regulatory bodies to inspect vehicles, sites and tipping records in the normal way to ensure compliance, this approach would oblige all waste collectors, authorities or municipalities to ensure the complete effective segregation of all the putrescible waste matter from all of the other components of MSW prior to arrival at the site. Given the current best available dirty MRF technology, as discussed, this level of efficiency is simply impossible to achieve from mixed waste inputs. Hence, a total ban of this kind would establish a simultaneous requirement for householder source separation, though, as we have seen, this would have significant benefits in respect of final product quality. The cost implications are harder to foresee, since, particularly in terms of haulage, much depends on how much biowaste could be treated locally, though the necessarily fast establishment of suitable facilities to deal with this additional load will itself come at a price. Despite the option's robust appeal, it is very unlikely that anywhere will seek to implement a total ban on biodegradable waste as a first move. It may become the logical final step at some point in the future, but that is a decision not likely to become relevant for quite some time.

Selective Ban

A more realistic possibility would be a selective ban, prohibiting the deposition of certain specific kinds of biowaste. This might be easier to implement, particularly since expressly 'biodegradable' material, certainly under the terms of the EU Directive, is, in the form of paper and card, currently one of the most commonly diverted fractions from the typical household waste stream. At a later date, such an approach would be flexible enough to allow the extension of the specified classes to include garden/yard waste as the required infrastructure and bio-treatment facilities are developed. The benefits in terms of the effect that such a partial ban, especially one predominantly based at the outset on paper, might have on methane emissions is harder to predict, since these materials, though biodegradable in the strictest sense, are less readily broken down than truly putrescible, fresh cellulosic waste. Likewise, the transport element is difficult to gauge, since current recycling arrangements often necessitate the haulage of paper over long distances to the mills and any overall increase in this would inevitably give rise to additional fossil fuel usage, air pollutants and greenhouse gases, all of which have their own environmental implications, particularly in the light of the Kyoto protocol.

Such an arrangement shares with the previous option the imposition of an across-the-board duty on operators and authorities to desist from landfilling the specified types of waste. Though it does not call for the same rigour of separation as a total ban, it does, nevertheless, impose a segregation obligation on all waste collectors in respect of these particular materials, which may, in a practical sense, prove every bit as difficult to achieve. Moreover, since it is highly likely that the selective ban would ultimately be extended to cover the more typically putrescible component of MSW, it is probable that this would also force a faster pace on the provision of alternative routes for dealing with this form of biowaste than would ideally be the case. Since the long-term aim of landfill diversion is to minimise the environmental implications of biodegradable material, it is clear that the real goal is more about finding other rational options for biowaste than about merely choosing a different disposal route, or simply cutting methane emissions, important though this is in itself. Consequently, though the concept of a partial ban equally has its merits, its limitations and inherent inflexibility mean it, also, is unlikely to be widely adopted as a means of promoting the alternative treatment of biological wastes.

Limiting Actual Amounts

One of the themes which seems to re-surface with depressing regularity in any waste management forum is the difficulty in obtaining truly verifiable and representative data on waste arisings. When constructing a strategy designed to control the influx of biowaste into landfill, particularly when such an instrument is obligately intended to guarantee compliance with statutory requirements, as in the case of the EU Directive, the absence of such information calls into question the manner by which the performance of that approach may be judged. Should the amount of biowaste entering municipal tips be inferred from the amount known to be diverted elsewhere, for alternative treatments, or directly, via gate receipts? The former gives a picture of the state of waste bio-cycling, the latter of the actual physical amount being buried, but neither provides a true measure of diversion. Since the aim of methane reduction is central to much of the thrust of biowaste exclusion, for reasons relevant to both waste management and climate change, it seems more sensible to concern ourselves with the actual amount consigned to landfill. It is, however, important to be aware that, though hopefully fairly closely related to it, this is not exactly the same thing as an authentically quantifiable diversionary total, though it may be the nearest we can realistically get to one. This leads on to a third possible means for controlling the putrescible input, as it then becomes possible to impose limits on the amount of this material which can be disposed of in this way. Such an approach could be couched in terms of a total overall tonnage of biodegradable waste permitted to enter landfill and thus might be itself aimed at either the waste company, restricting the amount it could accept, either generally, or specifically on a site by site basis, or on the local authority, controlling how much biodegradable material they could send for landfilling in the first place. While this could simply be based on an assumption of the average overall percentage made up by putrescible material in MSW, and the calculations made accordingly, this would seem to run counter to the larger intention of diversionary thinking. A method of more accurately reflecting a given area's true state of waste management would seem to be a better approach. This would have the added advantage of largely removing the necessity for huge additional householder segregation initiatives, since more precise estimates of the likely actual biowaste content of loads from given sources could be directly factored in to this model, enabling fair recompense to be given in respect of the relevant degree of prior separation. Thus, the need for the kind of extremely effective sorting of the putrescible component, implicitly required for both of the 'ban' options, is neatly side-stepped. However, it must also be borne in mind that particularly effective removal of one or more of the dry recyclable elements from the waste stream could have the effect of actually increasing the observed overall proportion of biowaste in the residual refuse.

The question as to on whom the burden should lie for regulating the use of landfill is of considerable importance, both for those directly involved and possibly for the future complexion of the waste management industry. There is a clear case to be made for placing the onus on the site operators themselves, since this enables them to retain control over the use of their facilities and to plan and invest for future provision accordingly. Maintenance of the *de facto* relationship between established commercial companies and local authorities has some merit, allowing the major decisions regarding the provision and design of services to rest with the operators themselves. Even so, the issue would still remain as to how the necessary allocation of resources would proceed on a national basis, if allowing a given landfill to accept waste in line with previous established demand is not to penalise those areas which have already begun to take steps to break their dependency. Clearly, permitting those areas with historically heavy landfill use to continue in the long-term, conflicts heavily with the wider objective. Moreover, it is by no means certain that if control effectively remains with the landfill operators that there will be enough freedom and flexibility for local authorities to be able to implement new initiatives or plan their future waste strategies. A system in which the municipalities themselves are expressly charged with oversight of the final destiny of biowaste sits comfortably with their role in planning and also with their overall responsibilities for waste in general. The main logical appeal of this is that it very effectively puts the decisions back into the local arena, which is where it most properly belongs. With greater local discretion and control over the management of waste, the likelihood is that there is considerably more chance of the truly best practicable environmental option (BPEO) being obtained. Certainly, the in-built flexibility of this approach, particularly in respect of a potentially locally sensitive consideration like waste management, has much to recommend it.

Tradable Permits

There has been a suggestion[1] that tradable permits, supplied to either the site operators or the local authorities themselves would help ensure the proper running of such options. It has been further proposed that this system would offer the greatest flexibility as it allows landfill usage in a given area to change in response to alterations in circumstances and local requirements over time, which may be a powerful argument given the expected increased availability of biological treatment facilities. Certainly, the ability to trade permits opens up the potential for the generation of extra transitional funding for those municipalities wishing to forge ahead with the early provision of their own biowaste plants. In this way, areas which have a particularly high need for landfill could purchase 'unwanted' permits from another authority or operator, enabling themselves to meet their immediate requirements and the sellers to offset or subsidise the cost of their own arrangements. At least in theory, this has some merit as a measure to kick-start the development of real biological waste treatment, though careful administration may be required to ensure that this actually happens. Previous experience in the UK with packaging recovery notes (PRNs), trading in which was intended to stimulate a growth in grass-roots recycling and recovery, shows that it is not always the intended beneficiary of such programmes who comes off best. In addition, if the permits are allocated to existing waste management companies, there is a danger that barriers to new entrants may be formed, which may affect both the provision of new sites and the rise of new companies. While this could, in the short term, have the effect of making the distribution of landfill outlets for biowaste somewhat uneven, the more worrying outcome might be that emerging companies, actively promoting alternative technologies for biological waste treatment could be disadvantaged by the permit system itself. There is, perhaps, more of a case for leaving the permits in the hands of the municipalities themselves, not least because this makes a good fit with their existing roles in the administration of planning and waste management. Also, the needs of local authorities with respect to biowaste will show inevitable changes over time as the raft of waste initiatives, including minimisation, packaging reduction and increased biological waste treatment itself, begin to take effect. A tradable permit system, if designed with sufficient in-built flexibility and specifically vested with the authorities themselves, would enable this kind of variability to be accommodated. Nevertheless, any such arrangement would still require some form of fair allocation at the outset, to ensure that the resulting burdens and benefits were properly apportioned. It would, therefore, be important to take into account a number of other factors, such as the population, including its age and socio-economic profiles, the location of all waste sites in the area, the actual amount of locally available landfill and the extent of accessible specialised biowaste treatment facilities. It might also prove necessary to consider wider issues of householder source segregation, additional available funding, a raising of the awareness of cost-based accountability in waste management and possibly the renegotiation of long-term contracts between private sector waste management companies and local authorities, before this could finally be achieved.

Developing Alternative Approaches

The second of the two areas in which action is required is the development of suitable biowaste options as alternatives to landfill dependency, which potentially include both biological treatment technologies and wider minimisation initiatives. There is considerable scope for the further extension of the technologies discussed earlier in this book, beyond the current level of use. Of these, currently the greatest contribution is made by composting in one of its guises, with anaerobic digestion close behind, in second place. How each of these will fare in the future depends largely on external influences like progress in mechanical separation methods, the wider uptake of householder source segregation and the available markets for the product. Since these are all essentially driven by local considerations themselves, it seems unlikely that there will ever be a single, standardised approach to the biowaste question. The development of biological treatments in the commercial waste management world of landfill and incineration is intimately linked to the development of sustainable markets and these are, themselves, dependent on consumer confidence. Many municipalities currently sell MSW-derived composts or humus products to their citizens, but they are in the position of being able to offset this against running costs or mitigate the expense otherwise incurred on their alternative disposal route. Few commercial companies will be able to approach biowaste in this way, and particularly not under circumstances of increased diversion and therefore more readily abundant rival products, of fundamentally similar character. The issue will, ultimately, rest on quality control and the guaranteed nature of the compost material, but it is equally clear that old-fashioned marketing will play a major role in establishing the general acceptance of biowaste-based soil amendments.

As alternatives to landfill, both AD and composting can be said to be already fairly well established in the wider consciousness and fit easily into the waste hierarchies which a number of countries have produced. Recently there has been a swing towards viewing the aims of reduce, reuse and recycle as more of a spectrum rather than a set of absolutes, which, certainly from the biowaste standpoint, seems to provide a better reflection of the situation. In any case, there is a danger that, while deep in semantic arguments over the distinction between re-using and re-cycling, the whole point of the exercise becomes lost. Irrespective of what name is given to the land use of a waste-derived soil amendment, that use remains beneficial. There is, then, a broad consensus of acceptance of the desirability of biowaste processing, to which the wider application of composting or digestion technologies is a logical extension. Of all of the current methods of biological waste treatment, these are the most likely to be front runners in areas making their first moves towards the provision of local facilities. Which of these techniques will be chosen will, like so much to do with biowaste, largely depend on local circumstance. In terms of the eventual marketability of the final product, as we have seen, both AD and composting operations make the best quality soil enhancers when the input feedstock is cleanest. Hence, if the sale of the product is essential to the viability of the scheme, then considerations of collection and separation are likely to be more important than the technology itself. However, specific characterisation of the feed material may have some relevance in deciding the latter.

One of the main factors in favour of composting, at least for what might be described as intermediate-scale operations, is the fact that such facilities are relatively easy to run. Thus, the local authority can effectively increase its waste management options without swelling costs, by operating the plant for itself. A simple windrow set-up demands little more in equipment than the council is likely already to possess and the routine turning and maintenance required can often be scheduled into the duties of the existing workforce, who are on the pay-roll from the outset. This kind of approach, typified by the treatment of garden waste commonly arising from the likes of a civic amenity site, has proved both popular and successful in many areas of the world and provides an important tool in the move towards greater landfill diversion. It is also, of course, an expressly local solution which has much relevance to what has been termed the 'proximity principle'. At its simplest, this dictates that waste should be dealt with as close as possible to the point at which it arises. This itself has important implications both at the immediately local level and beyond. For any consideration of self-sufficiency or sustainability, clearly a region's in-house ability to treat waste must be taken into consideration, but even more simply that, there should be a realisation amongst the general public that once they have thrown away their refuse, it does not simply cease to be a problem. In many respects, this is critical to any attempt at sustainable waste management for biowaste or, indeed, any other component material of MSW. At the larger scale, this principle ensures that one region's problem is not simply exported to another region or country and, moreover, it implicitly acknowledges the potential environmental damage involved in waste transport itself. A municipal composting centre, then, both contributes to the wider biowaste goal, while simultaneously making a very clear statement of local responsibility taken for waste management. Aside of their evident reinforcement of the central environmental role of the authority, such facilities become, almost without exception, sources of great pride to citizens and councils alike.

Although many operations at this scale will tend to make use of composting, the relative ease of management and running will often facilitate the use of other low intervention technologies, like vermiculture and may even encourage the development of sequential treatment trains, or even a truly integrated local approach. There is a natural fit between the biological processing of biowaste from households and municipal parks and the eventual return of the derived soil amendment produced to grow another year's flowers for the area's gardens, both public and private. The establishment of SRC for a local heating scheme, grown on land mulched with local biowaste-derived material is simply the logical extension of this. Coupling an approach of this kind with the collection of paper, which is a widespread and common local authority strategy, effectively provides an instant reduction in the two largest biodegradable elements in MSW. While this may be viewed by some as little more than a 'numbers game' in respect of the European situation, even slowly biodegradable substances like paper and card unquestionably do make a contribution to *in situ* methanogenesis. Though, as was discussed in the previous chapter, it has been suggested that burying this material in landfills could be a valuable means of carbon sequestration, ultimately there seems little value in a lock-up mechanism which decomposes over so long a period, so unpredictably and, in the end, pays back with dividends.

Taxation Issues

It is likely that more municipalities will adopt these measures, or other like them, to begin to divert increasing amounts of their biowaste. In many areas, there is an additional financial incentive in that a number of countries have implemented a tax on tipped waste, in order to begin to drive the establishment of viable alternatives to landfill. Such instruments may have different bandings applicable to the type of waste, as in the UK, where closely defined inert materials attract tax at a lower rate than 'active waste'. Some of those nations with landfill taxes have used them to establish a periodic escalator on tipping, making higher than inflation increases in the duty payable, in a deliberate attempt to channel waste elsewhere. However, the efficiency of such measures is not guaranteed. According to the findings of an inquiry begun in February 1999, by the House of Commons Environment Committee, raising the landfill tax by the kind of modest increments planned will have little effect on the amount of refuse using this route for disposal. When the tax was first imposed, in 1996, active waste was charged at a rate of £7 per tonne, a figure which rose to £10, in April 1998, with the intention being to increase it, in stages, to £15 per tonne by 2004. The House of Commons Committee, however, found that the original

£7 level had made little difference to landfill disposal and concluded that it was unlikely that the £10 tax rate would significantly advance waste minimisation or recycling. To produce the kind of incentive which would actively shift the balance sufficiently to make this happen, the Committee suggested that, by 2004, the amount due should be £20 per tonne, with continued rises thereafter to bring the levy up to £30.

The same study also highlighted one of the other potential drawbacks to the idea of this kind of taxation, at least as it applies to MSW. There is evidence to suggest that, far from stimulating the development of alternatives, the tax may actually be inhibiting it, since the budget elements previously used by some municipal waste managers to support diversion initiatives, are being swallowed up as payment contributions. Moreover, this effect is compounded by the increase in fly-tipping, which further adds to the financial and practical difficulties of the local authority. Much of the solution to this problem depends on the way in which the individual scheme is administered. In the UK, for instance, the regulations permit the site operators to use up to 20% of their liability under the scheme to fund environmental activities, chosen, at their discretion, from within particular guidelines, the remainder going to the public purse. Ensuring that more is allocated to projects actively designed to encourage alternative waste treatments in future, rather than to those offering less direct benefits, was one approach suggested by the Environment Committee. Another was to set a level for the amount of tax revenue going to the Treasury, channelling any surplus back into promoting reduction, reuse and recycling programmes. This would seem to be a particularly appealing form of natural justice, since, should the tax fail to deliver the desired reduction in landfill use, the proportion of the collected money earmarked for the development of these initiatives would rise accordingly.

This type of financial instrument also offers a potential solution to one of the problems with the kind of permit system discussed earlier, which places the control in the hands of the local authorities themselves. Under such an arrangement, there is an obvious difficulty in providing effective sanctions to make enforcement realistic. The instigation of a two-tier taxation system might, in the final analysis, be a good 'carrot and stick' approach to ensuring local authority compliance. Municipalities meeting recycling and diversion targets could be rewarded by permitting the waste that they do landfill to attract a lower rate of tax, while other authorities pay at the higher amount. This is a swift acting and direct approach to the problem, the rigours of which could be avoided simply by the achievement of the required goals, though a measure of flexibility in the setting of these might be necessary to allow for legitimate regional variances. The beauty of this tax, however it is set or administered, is that it is politically safe, since to object to it is to apparently condone the despoiling of the environment. Moreover, its value as a diversionary instrument lies in the fact that there are clear savings to be made if a suitable non-landfill option can be found and as budgets increasingly come under pressure, the value of excluding putrescible material likewise grows.

However, useful though the intermediate-scale approaches undoubtedly are, it seems unlikely, on viewing the scale of the final required diversion, that they will be sufficient alone. The UK government's preliminary estimates indicate that some 62.5% of MSW is 'biodegradable' under the Landfill Directive definition.[2] Taking the final target of a 65% reduction in putrescible waste required, this represents over 10 million tonnes of biowaste, just from the UK, to be diverted annually into alternative treatment routes and this calculation takes no account of the fact that total waste arisings appear to be growing by about 3% per year.[3] This is a very large quantity of material, amounting to something like a third of the country's overall waste stream. Consequently, there will be an inevitable requirement for increased centralised facilities, probably using composting or anaerobic digestion technologies in the main, though for some areas and sets of local circumstances, other approaches may be found to be appropriate. The existing methods of biological waste treatment are clearly suitable to meet these additional requirements, though obviously the scale of their provision will need to be significantly expanded. In many respects the major difficulty with their application lies not in their ability to process the biowaste, nor with the biowaste itself, but rather in the logistics of separating and rendering the putrescible fraction into a pure enough feedstock, on so large a scale. As has been discussed at length previously, the quality of the input material is the decisive factor in the eventual product character and, thus, in its acceptability to end-user markets. While there is undoubtedly scope for advances in biotechnology to produce new treatment methods, or improvements on the present ones, which may represent significant potential long-term opportunities, as in the case of the developments in biowaste ethanol fermentation mentioned in chapter 8, only a satisfactory resolution of the segregation issue will make any of it possible.

Both in terms of the application of biological treatments themselves and in respect of monitoring compliance with any established targets for diversion, it is essential to have an effective system of distinguishing between waste components. The difficulties inherent in the lack of an agreed standard classification of MSW elements, the practical problems in obtaining a truly representative refuse profile model and the scarcity of historical data from which to discern long-term trends have been highlighted in earlier chapters. Consequently, any moves towards attempting to provide the kind of necessary fractional separation required to meet the demands of the previous paragraph will need to retain a marked degree of flexibility, to deal with not only the discrepancies between the model and empirical reality, but also variations over time. Again, there is an inverse relationship in the equilibrium between the degree of segregation done by the householder, prior to collection, and the complexity of the centralised sorting activity required to finish the job, and this holds true, irrespective of whether the on-site separation is achieved by manual or mechanical means. Many established biowaste treatment initiatives have favoured a heavy dependence on the front end, with the material arriving at the facility already effectively process-ready. The cost element of this is hard to pin down, since it is very dependent on local conditions and economics. While this approach does away with operational expenditure on either additional sorting machinery or personnel, it may have cost implications in terms of the collection arrangements, and the financial balance between these two extremes can only be found by reference to factors like the local labour rates and fuel pricing. It is probably fair to say, however, that solely in respect of the derived fractional purity, segregation at source seems to give better and more consistent results.

For a number of countries, states and regions, separate collections are already a matter of course and it is, therefore, a relatively simple matter to make any small changes in such existing system which may be required to accommodate higher levels of biowaste treatment. The situation for those which have traditionally relied on landfill to meet the bulk of their waste management needs is more difficult, since this generally entails a major change, doing away from the established single collection of mixed

MSW and implementing an entirely new arrangement in its place. This may also affect any incinerator-dependent localities seeking to increase the extent of their biological waste processing. There seems to be a strong tendency towards conservatism in the public's views of waste management, at least as it directly affects them. In one very small and informal study, over three-quarters of the people asked, who routinely visited one particular UK recycling centre, were in favour of mandatory separate collections. This fell to under a third amongst a similar sample from the same area, who used the site only occasionally, to rid themselves of exceptional items, for which there was no other ready means of disposal.[4] It would be unfair to conclude too much from such a survey, but, while there is widespread general agreement with the aims of biological waste treatment, it would seem that there may be some reluctance to adopt personally inconvenient measures to bring it about, at least in the initial stages. Clearly, if this is the case, social change and the manner in which it is encouraged may prove to be every bit as important as the provision of appropriate facilities, the establishment of end-user outlets and the robustness of the treatment technologies themselves.

The Future of MRFs

As a result of these kinds of concerns there has been some interest over recent years in the dirty MRF concept. The main idea driving this was that, for areas which had historically featured a single mixed waste collection, rather than attempt to change the weight of traditional practice, it would be easier to allow the refuse to continue to be gathered in just the same way, and then separate it at the processing plant. While this would, inevitably, require greater investment in time, manpower and equipment by the operator, the intention was that this would be offset by the consequent greater degree of control achieved over the production of the feedstock and the easing of the transition to specialist biowaste processing. Furthermore, such a system was believed to able to circumvent one of the major problems of long-term waste provision planning, namely changes with time of the refuse itself, by adopting a modular approach to the devices used to sort the MSW. The reasoning was that as the waste character changed or advances in technology made better separation methods available, old components of the sorting line could be retired and replaced with more appropriate ones. In principle, this approach has much logical appeal, but the practical fulfilment of the dream seems not to have proved possible, largely for reasons discussed elsewhere in this book. As well as underestimating the absolute importance of feedstock purity to the eventual product, advocates of the dirty MRF may also have failed to take into account the likely topography of future waste management arrangements to deal most effectively with an increased requirement for separate biowaste treatment. The idea of a mechanical sorting plant fits best into a centralised, single-site scenario, to which all the MSW collected comes for separation and subsequent processing. In this vision, the operator of the MRF also runs the recycling, bio-treatment and disposal initiatives as a single unit. While integration within waste management options is a major goal, there is a strong body of evidence, based on current best practice around the world, that seems to suggest that the one-stop approach seldom happens in reality. Indeed, there are sound arguments to be made that the individual component parts of any such holistic overall scheme should be separate. This enables the maintenance of adequate business competitiveness as well as meeting the more directly environmental concern to ensure that those with appropriate expertise, rather than generalists, run the specialist elements.

There is a growing feeling that the future role of the MRF may not be in the separation of mixed MSW itself, but rather to cope with sorting the various recoverable materials from what have been termed 'single stream' recycling schemes.[5] Conventional wisdom has always held that effective recovery rates depend on the maximisation of quality achieved by source separation coupled with the simultaneous drive towards ever greater throughput of material. Some initiatives have required very detailed segregation of the recyclables, often demanding separate categorisation of the different glass colours and types of paper. As has been mentioned previously, the more that is done at the front of the recovery process, the less cost is involved in the processing itself and, equally importantly, the greater unit value the reclaimed materials have in the secondary markets. However, there is a rising tendency to challenge some of the fundamental premises of this approach to recycling, particularly, as has also been discussed earlier, the tacit assumption of a universal willingness to participate. It had always been assumed that, as increasing numbers took part, and diversion rates rose, the marginal cost of collections would slowly fall. There is a tendency for this not to happen, with participation typically tailing off, to form a background plateau rate, which leaves collection costs higher than anticipated and the promised massive economies of scale are not achieved, since the necessary volumes of materials do not appear. Hence, the single stream approach stands as a sort of 'half-way house' between these traditional highly segregated separate collections and true mixed MSW. There may be some value, as examined in earlier chapters, for the use of dirty MRFs in those areas which make mixed waste collections, to provide a coarse biowaste-enriched fraction suitable for some form of crude biological treatment, thus forming a bridge to ease the transition between current and future practices. It is likely that the first step along that route for many authorities will be the establishment of source segregation of waste into the two simple categories of biodegradable and mixed recyclable materials. Such householder separation of the former is likely to be, at the very least, every bit as good as that capable of being achieved by mechanical means and, in all probability, significantly better. The sorting of the latter into the required material types will then become the preserve of the MRF, an approach which is not without merit, not least since a machine is easier to reprogramme in the event that secondary markets dictate different categorisations in the future. It is, perhaps then, no surprise that Ademe, the French Agency for Environment and Energy Management, has predicted a 14% growth in France's domestic waste separation sector beyond 2000.[6]

Waste Minimisation

Waste minimisation initiatives may have an effect on the overall biowaste situation, but often in ways which are quite different from what might initially be expected. The main idea behind these drives lies at the most favoured end of the typical hierarchy; an attempt to reduce the amount of refuse produced in the first place. By cutting down the total waste arising, the quantities of material needing to be dealt with falls. For industry, once waste production is characterised and understood, there are a number of benefits to be gained from adopting this approach, principally in areas such as lower raw materials costs, increased efficiency in resource utilisation and reduced expenditure on waste disposal. Unsurprisingly, once the realisation dawns that there are savings to be made, even in the absence of a specific statutory obligation to implement minimisation, there is evidence of enthusiastic participation by businesses.[7] The situation with regard to household waste is not as straightforward, requiring major co-operation between government, manufacturers, retailers and the public to have any effect. It remains an area around which it is difficult to legislate, inevitably having to rely on the voluntary decisions of individual householders to take part. Applying a definition of *biodegradable* which includes paper and card, there is considerable scope for minimisation schemes to bring down the overall volumes thrown away. These materials lend themselves to control and limitation under measures as diverse as those intended to reduce excess packaging or aimed at cutting down on superfluous photocopying. However, since much of what might best be described as the 'fresh' biowaste found in MSW is food waste, it is difficult to see how any such reduction drive is likely to have an effect on the amount of this material present in household refuse. In any case, to persuade people away from highly packaged, instant meals is expressly to encourage them back to food in its original form and hence, while driving down one biodegradable component, necessarily increasing another. Where this particular balance point lies would be hard to determine and, in the final analysis, what really matters is that a suitable process is chosen to treat the residual biowaste, whatever its eventual character may be.

One way in which waste minimisation can have an unexpected effect is by skewing the overall compositional character of a given waste stream, which may have possible repercussions for long-term provision modelling, particularly if appropriate corrections are not made to compensate. It is clear that the kinds of waste which respond best to reduction are those materials which commonly arrive in the home as containers or packaging for the commodities which were actually wanted. Thus, it is possible, by avoiding excessively packaged goods, using shopping bags rather than free supermarket carriers, purchasing loose vegetables and opting for returnable containers rather than disposable ones, to make considerable difference to any given household's refuse output. It has been reported that some dedicated minimisers have managed an overall reduction in their waste production of 50% or more[8], though, obviously, it requires commitment, time and effort to achieve, and at the sort of level that it is clear not everyone would be willing, or indeed able, to match. Even at more typical rates of participation, such schemes, coupled with initiatives to boost other ways of reducing outright disposal, like the provision of recycling bins for paper, glass or cans, can bring about some measure of alteration in both the overall total waste stream tonnage and the relative proportion each waste type represents within it. For a given mixed MSW collection, characterised originally as a whole, the removal of significant amounts of individual waste materials, for whatever reason, will tend to make the remaining component elements appear to contribute a higher percentage of the new, lessened total, than previously. Although in this scenario, the overall quantity of refuse goes down, the relative fractional balance shifts. Such a change in the stream profile could have implications for any diversionary strategy judged on average load characteristics, like those discussed earlier in the chapter. However, waste minimisation, as it applies in the context of MSW, is realistically more likely to bring about reductions which are significant only at the individual household level, rather than causing a major swing in national waste figures and it is hard to imagine that this situation is likely to change in the future.

Sustainable Waste Management

What waste minimisation does begin to address is the issue of sustainability, which has assumed greater significance over recent years, and its influence seems set to continue to gain importance in future formulations of policy. The idea of sustainable development was defined by the Bruntland Commission back in 1987, under the aegis of the World Commission on Environment and Development, as an approach which 'meets the needs of the present without compromising the ability of future generations to meet their own needs'. This exposition of the principle has received widespread international acceptance. The main aims of its application have been further set out as social progress to address the requirements of all, effective environmental stewardship, the maintenance of high and stable economic growth and levels of employment, and the utilisation of natural resources in a prudent fashion.[9] Unsurprisingly, since moves of this kind, particularly in the latter two categories, tend to convey good commercial benefits, businesses have not been slow to see the value of this approach. According to a recent survey[10] of nearly five hundred environmental, health and safety and other business executives in North America and Europe, 95% reported that they felt sustainable development to be 'important' and a little over 80% further said that they thought it could be of significant real business value. Around 70% of European companies and more than 55% of those in the US were found to be actively utilising a sustainable development approach to strategy and operations across the organisation, seeking thus to gain business advantage, though most respondents reported that these initiatives were still only in their early stages. There seemed to be a number of factors perceived to be of value in this respect, including increased efficiency, competitive streamlining and better public relations. The contributions of technological innovation, work-force awareness and increasing customer expectations were also found to be of universally recognised significance in driving sustainable development forward.

Although strictly speaking, the conceptual division of sustainability, which must form a fundamentally unified approach to the management of all resources, is a logical impossibility, there is some merit in looking at a more keenly defined sector target, which we might thus term 'sustainable waste management'. This may be simply viewed as the use of materials more rationally and efficiently, so that wastage is reduced wherever possible and where it is not, managing the unavoidable waste produced in such a way that it contributes to, rather than conflicts with, the greater environmental, social and economic objectives of sustainable development. Interestingly, though the relevant European Commission Directives have been interpreted as 'working towards sustainability'[11] there is no firm EU definition of the idea of sustainable waste management itself. Globally, much of all waste arises as a result of the way resources are initially managed to produce and deliver goods and services and most of the remaining refuse is generated as a result of their subsequent use by consumers. The transition to seeing the potential of what is currently viewed solely as 'waste' as a useful raw material is a vital turning point in the move towards its future sustainable management. In the wider context of all wastes, simply disposing of a product is frequently a missed opportunity, since many can be re-used either for their original purpose or otherwise, or have value reclaimed from them in other ways, typically in terms of materials or energy. For the specific example of biowaste, of course, it is this latter approach which is relevant and in essence all forms of biological waste treatment revolve around either regaining useful material benefits, like the nutrient value and humus components, or using the organic material to derive energy by some means. Under the kind of short rotation coppicing regime previously discussed, it is even possible to go some way towards doing both. In this respect, the situation for biowaste recycling is better than for many other materials. With the bulk of conventional dry recyclables, the real desired upsurge in their value is linked to their increased utilisation in products which will themselves be resold, thus driving further reclamation by the market economy. Hence, before this can happen, there must be a willingness on the part of consumers to accept goods made from these secondary materials and, probably, a need for these products to show a price advantage over their virgin counterparts, particularly if the quality of the truly new is, or is perceived to be, superior. In the early stages of their production, a cheaper recycled product is, in itself, unlikely, as economies of scale will favour the established production method over the emergent. While, as has been discussed at length in earlier chapters, the direct sales of biowaste-derived soil products are also dependent on market and quality issues, there is already a general awareness and acceptance of compost, mulches and the like. Thus, at least the idea of a horticultural or landscape product made by a process of controlled rotting is not likely to be seen as something new or unproven and, though there may still be some reservations regarding the feedstock, the principle is well established. How well the operators of biowaste treatment sites will prove able to capitalise on this is, however, a different matter.

Best Practicable Environmental Option

There has been growing consideration of the relevance of the best practicable environmental option (BPEO) as applied to waste, not least because the concept itself encapsulates three of the four major goals of sustainable development and, thus, of sustainable waste management. The BPEO is defined as 'the outcome of a systematic and consultative decision making procedure which emphasises the protection and conservation of the environment across land, air and water . . . for a given set of objectives, the option provides the most benefits or the least damage to the environment as a whole, at acceptable cost, in the long term as well as in the short term.'[12] The application of this principle is a useful objective tool in the decision making process particularly in respect of the provision of biowaste facilities, though it can sometimes lead to conclusions which may, on the face of it, seem surprising. However, the rigour of these outcomes obtained under its strictest operation may be mitigated and further modified to fit the particular circumstances by reference to the supporting concepts of a waste hierarchy and the proximity principle.

The former enjoys widespread familiarity in a number of fundamentally similar forms, providing a useful conceptual framework which allows the relative merits of various possible refuse management options to be evaluated. In general terms, modern waste hierarchies tend to consider recycling as occupying a higher position than energy recovery and, thus, to view methods of biological waste treatment, like composting and anaerobic digestion, as falling within this definition of recycling. While this has obvious implications for the manufacture of soil products from biowaste, it must be emphasised that the 'energy recovery,' taken to be of lower priority, refers here to indiscriminate, mass-burn incineration, rather than the applications of specialist biotechnologies to the production of secondary fuels. The final criterion must be one of valid use, as opposed to convenient disposal. Direct combustion with energy recovery will always have its place for certain kinds of material which are 'biodegradable' in the strictest sense of the word; it cannot, however, truly be justified as a route for all biowaste.

The proximity principle, which was mentioned earlier in this chapter in respect of intermediate-scale composting facilities, forms the link between the more absolutist measures of the preceding concepts. In this way, the BPEO for a given waste material may be judged to be a method relatively lower in the hierarchy than might have been expected because the costs, either financial or environmental, of transporting it to a distant reprocessing plant are greater than the benefits to be derived from its recovery. In this way, though the environmental implications of the specific haulage option used must be taken into account, and not a mere consideration of the mileage, the case for consigning low-value materials over large distances for recycling is weak, set against a backdrop of high road haulage and heavy diesel use. As was pointed out in the previous chapter, this is one of the major arguments which has been advanced in favour of the combustion of 'biodegradable' paper for its energy value, rather than its continued use for recycling. This, in turn, leads us towards understanding both the necessity of sustainable waste management and, ultimately, of sustainability in the wider context.

The expression 'integrated waste management' has been frequently used and at its heart lies the way to the most effective strategy for all refuse in general and, thus for biowaste in particular. It has been suggested[13] that there are three basic key elements to an integrated approach. Firstly, it must clearly recognise, *a priori*, the holistic relevance of each step in the process. Thus any management decisions must view the collection, sorting, transport, processing and eventual product marketing or disposal as constituent elements of a single unified programme. Secondly, it must examine and exploit the potential influences and contributions to be made by all of the relevant parties in the waste chain. Thirdly, the actual management methods employed should be expressly designed to avoid over-dependence on any one technique and those options which are selected should be capable of working as near-seamlessly with each other as possible. In the context of biowaste management, the goals of integration are entirely consistent with the kinds of measures necessary to improve the rates of landfill diversion and the wider promotion of biological waste treatment. It is highly unlikely that any one approach will represent the BPEO for each and every one of the component elements in any given waste stream. In the same way, there will be scope within biological waste treatment for the applications of distinct technologies, either to deal with the different sub-fractions of biowaste, or as alternate methods to process the same material. Implicit within integration is the idea of flexibility, since it is the aggregate synthesis effect which matters, rather than the individual contributions which achieve it. This position is strengthened by further application of the proximity principle, as a series of complementary local strategies to treat the biowaste arising in any authority's area will be likely to prove preferable to an imposed and distant solution, even if such an approach delivers other benefits.

Integrated Resources Management

In many respects the big question is about sustainability itself, rather than simply the sustainable waste management element of it, though this does form a part of the greater whole and these aspects are, in truth, indivisible. In order to begin to understand how refuse treatment can be rationalised, it is necessary to examine the wider issues of resource management, which implicitly covers materials, waste and energy. The three are interlinked at so many different levels that it becomes impossible to consider them in isolation. If refuse, in general, is to come to be considered as a raw-material-in-waiting, then the bridge between materials and waste is clear. The cross-over between waste and energy currently means incineration and, though the combustion of some unwanted materials will always have a role to play, the present situation is a less than ideal fit. This is largely because any 'default option' burning misses the chance to make best use of the first bridge described above, by allowing little or no opportunity for reclamation. When this is extended to our larger environmental targets in terms of CO_2 reductions and fossil fuel usage, the position of biowaste becomes pivotal in the sustainability debate. If we cannot make our most basic and natural form of waste contribute to the cause of sustainable development, by reclaiming nutrients and other materials or to help provide suitably carbon-neutral fuels, then how can we hope to do the same for our artificial wastes? There are those who have sought to vilify the very idea of a self-sustaining civilisation by suggesting that the logical conclusion of the argument leaves us living in mud huts, deprived of all the benefits of science and technology. Assuming for a moment that this were to be so, we must still eat, and without a rational way of dealing with the biowaste that guarantees, even this *reductio ad absurdum* vision of the future fails to achieve what it sets out to do. There is, moreover, something of a numbers game to be played here, for if we adopt the Landfill Directive definition of 'biodegradable', then biowaste initiatives have relevance to around 60% of the typical waste stream. Thus, they have an opportunity to have a direct impact on a single fraction which is larger than all the other MSW component material streams put together.

In the final analysis, the future treatment of biowaste, however that is achieved, has widespread implications for the environment, the economy, society in general and human health. The way ahead, then, must face up to this and chart a course which, if it cannot do the most good, must do the least harm. Landfill sites release methane, over a lengthy period and in a largely uncontrolled way, but this gas can be captured and used to reduce fossil fuel usage, while being converted into a less potent greenhouse gas in the process. Energy from waste plants, including pyrolysis, gasification, fermentation and even anaerobic digestion facilities, not just incinerators, ultimately produce carbon dioxide, but theirs is 'current' CO_2 and may likewise help reduce the release of the 'fossil' form. Composting also contributes carbon to the atmosphere. Traditionally, incinerators and landfill sites have been commonly believed to pose some potential hazards to human health, and there is undoubtedly some evidence to support that position, in certain cases.[14] Biowaste treatment plants, though, along with the likes of transfer stations and materials recovery facilities, are not without their own potential health threats. No waste management option is entirely without its problems.

All biowaste, indeed all waste, makes a negative environmental contribution by virtue of its huge global transport requirement; thus the proximity principle itself stands as a clear example of doing the least harm, since to do the most good would be impractical, or perhaps that should read, *impracticable*. In this sense, it is not a 'technology' issue. There are many available methods to treat biowaste and it seems highly probable that more will be developed, and improvements made to these existing ones, in the future. Whatever approach to the overall goal of biowaste diversion is adopted, and whoever is made responsible for ensuring its smooth running, there will be room for a mixture of the individual technologies even within a single region, if that is desired. The biological basis of the method is, in some respects, relatively unimportant; what truly matters and, by the same token, will hold the key to success, is the manner of implementation. Political will and fiscal conditions form one part of this, by setting the scene, but finally, viability will depend upon whether the route chosen adequately addresses the fundamental practicalities necessary. With a good approach to sorting providing a clean feedstock and a ready outlet lined up for the final product (or energy) derived, the way is clear for any of the biological waste treatment technologies mentioned in this book to do the job. However, deciding which of these will perform this function the best is, like the selection of the prerequisite sorting and marketing options themselves, a matter for careful consideration, and not least of the many local factors which can exert their influence. In this respect, there may be many ways ahead.

References

1. *Limiting Landfill, A Consultation Paper on Limiting Landfill to Meet the EC Landfill Directive's Targets for the Landfill of Biodegradable Municipal Waste*, Department of the Environment, Transport and the Regions, October, 1999.
2. Ibid, p. 13
3. Ibid.
4. Author's own unpublished data.
5. *MRF Designs Around Single Stream Recycling*, BioCycle, August 1998, p. 45.
6. Environmental Data Interactive Exchange (EDIE) News Summary, www.edie.net.
7. *Cutting Your Losses: A Further Guide to Waste Minimisation for Businesses*, Department of Trade and Industry, 1992.
8. *An Introduction to Household Waste Management*, ETSU for the Department of Trade and Industry, 1998.
9. *A Way with Waste: A Draft Waste Strategy for England and Wales*, Part 1, Department of the Environment, Transport and the Regions, June 1999, p. 9, citing the UK Government's Sustainable Development Strategy, *A Better Quality of Life*, launched by the Rt. Hon. John Prescott MP, Deputy Prime Minister, in May of the same year.
10. Environmental Data Interactive Exchange (EDIE) News Summary, www.edie.net reporting a survey undertaken by the management consultancy Arthur D Little.
11. Dumble, P. and Whittaker, D., *Towards Sustainable Waste Management?* Proceedings of the Institute of Wastes Management, July, 1998, p. 23.
12. Royal Commission on Environmental Pollution, 12th Report, February 1988.
13. *A Way with Waste*, Part 1, p. 18.
14. Dolk H., Vrijheid M., Armstrong B., Abramsky L., Bianchi F., Garne E., Nelen V., Robert E., Scott J.E.S., Stone D., and Tenconi R., *Risk of Congenital Anomalies Near Hazardous-waste Landfill Sites in Europe: The EUROHAZCON Study*, The Lancet, Volume 352, Number 9126, 8th August, 1998.

CHAPTER 10
Policy and Planning

The Biowaste Treatment Plan

As was discussed in the previous chapter, there are a number of ways in which the objectives of increased biowaste diversion and its subsequent productive biological treatment can be achieved, both in terms of the overall policy implemented and the specific individual technologies employed. However, to ensure that the most appropriate methods are chosen for any particular waste stream and locality, there are a number of considerations which must be taken into account in the formulation of the biowaste treatment plan. The ultimate success of the scheme depends to a very large extent on how well this is drawn up in the first place and, subsequently, how well it is eventually executed in practice. The relevant factors can be divided into three groups, namely resources, procedures and politics.

The 'resources' heading is of paramount importance as it covers the three major elements of waste availability, capital and the workforce. Elsewhere in this book, the issue of waste availability has been discussed at length and it is an obvious, but necessary, point to reiterate that the nature and accessibility of a supply of suitable biowaste has enormous ramifications for the whole treatment plan. In certain areas it may be either necessary or beneficial to consider the entire range of available biological-origin waste from industrial and commercial sources, as well as the fraction available from MSW, in designing the processing programme. In the same way, the possible likelihood of demographic, social or other changes within the waste, over the project lifetime which the plan is intended to cover, also needs to be appraised and some attempt made to factor this into the overall decision making procedure. It may also be useful for what might be described as associated benefits to feature in the examination at this stage, like the potential for reductions in pollution, public health concerns or other forms of nuisance, if and where applicable.

The available capital is, likewise, an important resource to consider. Even if a given project is certain to become commercially viable in its own right, or is guaranteed adequate funding from municipal or other sources, the initial construction of the plant itself and the associated necessary infrastructure, together with any ancillary sorting or collection equipment, will require capital expenditure. Obtaining best value for money, in respect of the project costs against expected deliverable outcomes, obviously requires careful examination to be given to the use of invariably limited financial resources. While the initiative may well become ultimately self-financing by means of a combination of gate-fees, fiscal incentives and product sales, the up-front available capital defines the scope of the initial project impact and is generally fixed and immutable. The workforce element, on the other hand, both represents a current resource, as well as one which can be increased in the future to meet changing needs. Various skill levels are required for the implementation of the plan and the establishment of the programme, and these are principally determined by the treatment technology finally chosen. However, existing in-house expertise or specialist knowledge may be judged to be of sufficient worth that this may, itself, become a significant factor in the selection process for the method to be used, though such circumstances would normally be exceptional, as it is important not to let the tail wag the dog. Consideration should also be given to any likely expansion or development of the scheme expected over its lifetime, so that the relevant provisions for future manning levels and required skills may be included in the overall biowaste treatment plan.

The second major category, procedures, encompasses not only the details of the biological processing technology and its specific application to the project, but also the collection, handling, transport and final marketing of the materials involved. It is clear that to achieve maximum benefits and minimise the costs, both environmental and financial, from the project, the approaches selected to fulfil this section of the plan need to be efficient. Efficiency, however, may be only a relative term and depend, in part, on local circumstance. Thus, what is deemed acceptable in respect of biowaste collection and transport arrangements to one community, might be seen as overly draconian, or hopelessly lax, to others, depending on their currently established situation. There are many issues to take into account and a balance found between the environmental and the economic. Most of the biowaste processing technologies make essentially similar demands of the input feedstock. As has been discussed throughout this book, the character and relative purity of the material entering the biological treatment, has enormous bearing on the final product derived, irrespective of the process itself. In many respects, this is of significant advantage in the construction of a biowaste treatment plan. Since the input requirements are, to all intents and purposes, the same, no matter what system is favoured, there is no direct obligation to have first settled on the biotechnology before choosing the transport, handling and separation methods, though, of course, this does often happen. The eventual preparation details, like particle size, shredder outputs and final handling requirements will be contingent on the treatment approach, but the major logistics element of the procedural considerations are, thus, effectively independent of it.

Where *a priori* knowledge of the processing technology is essential is in the development of markets and marketing strategies. Obviously, in order to explore the available outlets for a product it is necessary to have a reasonable degree of assurance as to the likely nature of the product itself, particularly in respect of a material like a biowaste-derived soil amendment and especially in the early days of production. In the long term, public acceptance may increase on its own and a recognised standard may be developed, but even then, it is hard to see how a secure market could be developed for a product without some guarantee of its quality. The need for substantial and stable outlet, or in the last extreme, disposal, routes for the materials produced should feature heavily and early in the project planning cycle, for obvious reasons.

Political will, the final category, is a vital ingredient in any recycling scheme and is particularly important for biowaste reclamation and processing, since the intrinsic value is not as immediately apparent to the general public as it is for traditional recyclables like aluminium or glass. The role of government, both central and local, in legislating and determining how environmental initiatives will be translated into action, is clear, but this does not cover the wider aspects of their influence in educational programmes to promote public understanding, break down traditional barriers, remove uncertainties and debunk myths. This is a continuous process and without the political will to achieve it, progress for biological waste treatment, and the consequent success of any local project, will prove very hard to attain.

Objectives and Constraints

All of these factors are interwoven and thus, each influences the other to a greater or lesser extent, which in turn, leads to the overall treatment plan, which ultimately distils out from the consideration process, being generally more locally specific, than a mere sum of all its parts. It is, therefore, important that alongside a strategic review of the issues mentioned before, both the plan's objectives and constraints are clearly identified. Obviously, the main objectives of the programme are usually fairly self-selecting in terms of what can be achieved and the general benefits which will accrue from doing so. Potential pollution control avenues, environmental benefits and cost savings are typically readily defined in the early stages, as these are 'big-picture' aims and tend to be based heavily in the universally accepted eco-criteria which they encompass. The specific arena in which a plan's objectives are less frequently so clear-cut, or tightly thought through, especially in the initial phases of planning, relates to the total duration of the project, the actual geographical area and the particular waste streams involved. These aspects are of relevance to all recycling attempts, but they are particularly germane to biowaste initiatives.

Many aspects of the whole biowaste treatment plan are dictated by the intended length of the project's life. Local circumstances will play a large part in establishing the likely duration, which is also linked to the availability of resources and to longer-term scheduled aims. For the purposes of technology evaluation, which features later in the overall planning process, it is essential that this time-objective has been realistically examined and defined, since many of the decisions to be made at that point will necessarily be contingent on the robustness and projected lifespan of the systems to be employed. In addition, the initiative's duration may have further implications for the method selection process, since the longer the intended life-time, the more exposed the programme becomes to the effect of natural variations in the waste arisings themselves.

A consideration of the scheme's regional extent can be a very important factor in its overall success; the 'where' of waste collection can be every bit as critical as the 'how' and, indeed, in some instances, this alone may determine the whole method used. The physical situation of the plant itself may also have a major bearing on its usefulness, both to the individual municipality and for any desired potential expansion to local, or regional, facility status. Adequate weighing of these long-term objectives as the plan is drawn up can prove to be highly beneficial to the project's overall performance. Closely linked to the question of the area involved in the project, is the matter of the available biowaste itself. Particularly when it is intended to provide facilities for treating commercial or industrial biological origin waste alongside that arising from MSW, and obtain the benefits which attach to such economies of scale, where the plant is located influences how much, and what kind, of material is taken in for processing. It is equally possible that the presence of an existing food-processing plant or other large-scale biowaste source may, itself, be a persuasive consideration in the siting of the biological treatment facility.

There are a variety of potential constraints on biowaste initiatives which can often act as limiting factors on the instigation and development of biological waste treatment. Early identification of these may save a considerable amount of unnecessary work later, as well as tending to narrow down the viable options entering the eventual technology selection phase. Some of these are related to aspects of the resources and political categories already discussed. Limitations of available finance and skilled labour may be major problems, particularly in the case of capital-intensive proposals or highly biotechnical approaches. It is, obviously, vital to establish the relative influence of these at the outset, since their effect on the wider project goals can be of prime importance. Site constraints are another significant area which must be taken into account, whether or not an individual parcel of land has already been ear-marked to house the project. A municipality may be fortunate enough to enjoy the luxury of placing its plant ideally, but often it will have little or no choice in where it has to be located, since land availability, position, the location of other key topographic features and proximity to the major waste producers may severely restrict the options open for development. Even when this issue has been resolved, and all existing land uses and precedents have been borne in mind, specific environmental constraints may still feature in the overall equation. Particular aspects of the site's geology and hydrogeology, together with the current usage of the surrounding area and the likely impact on neighbouring industry, or housing, may need to be considered further and, ultimately, may have a large influence on the permissible type of biological treatment technology finally chosen.

Another factor which may act to constrain the implementation of biowaste processing is the question of public acceptability, which has some overlap with the political considerations mentioned earlier in the chapter. At one end of the scale, this encompasses harnessing the genuine goodwill towards rational recycling and waste management initiatives often encountered from the general public. At the other, it means sensitively and constructively circumventing natural suspicion, especially amongst those living near to the proposed facility. This is a particularly important aspect for biowaste projects, since the relatively low value of the products typically produced may rely on a sympathetic population to provide an outlet, if not an actual market. There has been much public disquiet in many countries, and at different times, over various health, and other concerns, arising from refuse-related issues, especially in respect of landfill and incineration. Any biowaste treatment plan must give recognition to the potential for this sort of local unease, not least since it enables the active promotional measures and on-going education process, mentioned previously, to be used to their best effect. It is an obvious, but sadly, often forgotten, fact that much of how any given proposal is viewed depends on the manner in which it is initially presented. There have been occasions when fundamentally good initiatives have failed to gain widespread initial support, through simple misunderstandings or misconceptions regarding certain of their details and intentions. Careful consideration of these issues at the outset of the project planning stage can help to mitigate much of the constraining effect of public opinion.

A final factor which may limit the rate of progress in formulating the overall plan is the commonly encountered lack of adequately detailed waste data. The decisions ultimately to be made regarding technology selection may require a greater or lesser degree of waste characterisation to be possible and, as discussed in chapter 2, frequently the picture is an incomplete one. It may, thus, be necessary for this information shortfall to be made up in the opening stages of the project, or even as a preliminary phase, to be completed before the rest of the work begins. While knowledge of biowaste availability delineates the project's potential feasibility, it is an awareness of that waste's compositional profile that determines the practicality of the proposal, not least since this helps define the likely market outlets for the final product.

Technology Selection

The selection of the actual biological treatment process to be employed is subject to two sets of influences which will be termed 'intrinsic' and 'external' for the purposes of the present discussion. Thus, an intrinsic factor is one which relates to an aspect of the intended technology or its application, arising by virtue of some property of the waste, or of the process itself, and which would always be a consideration, irrespective of external circumstance. The 'externals' are outside influences, the effects of which depend on local conditions, and are specific to the individual project. It should be obvious, therefore, that, particularly where a number of options are being considered for implementation, objective evaluation will necessitate each to be examined across a wide spectrum of criteria, to ensure arriving at the one which will, hopefully, perform best in any given instance. In this way, aspects of financing arrangements, running costs, environmental benefits, public concerns and local priorities may have a bearing on the decision. The development of a defensible scheme to quantify and compare, for example, the relative merits of hard-cash against environmental gains, will be essential, but this itself is unlikely to be transferrable to other situations, since it must, obligately, take account of the local external factors alongside the intrinsics. The simple device of making these comparisons on a per tonne basis has been widely advocated and certainly this approach enables a ready snap-shot of both the present and possible future implications to be obtained, which may be of considerable relevance to those hoping to drive a major expansion of the biowaste programme over time.

As a final general point on this issue, the consideration of any technology originally developed in a different country needs particular care, since there is considerable scope for a mis-match between the imported method and the relevant local external factors. While excellent systems for the collection, separation or processing of biowaste have been produced independently, in many countries of the world, it can be a mistake to imagine that they can simply be transplanted wholesale into a foreign setting. It may be possible to adapt them, of course, but specialist methods designed for the rural areas of the Mediterranean, for example, seldom adapt well to use in the colder industrial cities of northern Europe. Moreover, the very aspects which make them work in their own home may be entirely absent elsewhere. An American dirty MRF, for instance, designed around a refuse stream from which much of the food waste has been removed by in-sink disposal units, fares poorly when faced with traditional British mixed MSW, in which this biowaste is very clearly present.

Intrinsic Factors

Of all the intrinsic factors, the single most important is that the technology being considered must be capable of dealing with the specific type of biowaste available. Although generally any biological treatment can be beneficially applied to the biodegradable component of waste, for particular mixes, some individual methods may be better suited than others and this might form a logical limit on the usefulness of a given approach. This reinforces the need for the adequate waste characterisation data discussed earlier and also highlights the benefits of acquiring comprehensive information about the particular method proposed, rather than simply settling for a generalised overview of the relevant technology sector. Not all forms of composting or AD are the same and while, clearly, a broad grasp of the biotechnologies involved is helpful, in the selection phase of the plan, there can be no substitute for process-specific details. Having effectively looked at the front-end, by considering the input feedstock, the next focus of attention should be the back of the process, since there is no point in deriving a final product for which there is no likely market or use. This is not a problem in the case of most of the established technologies, but it may be significant for some of the others, and though this is strictly an intrinsic factor under the earlier definition, it will be returned to later, since its wider implications are better examined amid contributory external influences.

Process safety, together with issues of product consistency and quality, must also be taken into consideration, since they are fundamental to the success of the scheme. Though the particular statutory requirements for these may vary regionally, in absolute terms, the best that can be achieved for each arises as a result of the nature of the individual technology, the biowaste input and the manner in which they interface with each other. Thus, they may truly be considered intrinsic factors and their performance clearly relates back to the original matter of ensuring the fundamental suitability of any given approach to the particular waste stream. This is an important element in deriving the biowaste treatment plan, for, although the nature of the available putrescible refuse is subject to the normal variations over time, it is effectively the fixed element in the equation. It is important to view the selection of the treatment technology from this standpoint, rather than to make a decision based on the perceived benefits of the process alone. In many cases, both of these approaches will ultimately lead to the same result, but where there is any uncertainty or disparity, the needs defined by the biowaste profile must be accorded the greater significance. From earlier discussions it should be obvious that the quality of the final product has significant bearing on project viability and, like the issue of market readiness raised in the previous paragraph, this is an aspect which also has its implications at a more local level.

Another method-specific factor often given a relatively high weighting in the overall process of technology comparison is the proven nature of each. Unsurprisingly, local authorities often show a marked inclination towards methods which have long or successful track-records. While this may not always be an ideal measure, particularly if it relates to use on different waste types or was attained abroad, it does at least demonstrate that the approach can be made to work. The situation for newly developed biowaste treatments is, obviously, more difficult in the light of any such tendency towards an unwillingness to take a risk on their uncertain commercial efficacy, though it is an entirely understandable stance. This is particularly problematic for novel technologies which are, or even simply appear to be, radically different from established methods and especially when the applications for which they are being suggested are large-scale. For such approaches it seems likely that confidence building via a series of smaller successes will remain an essential prerequisite prior to their acceptance for consideration for bigger projects. As a general guideline for selection, given the choice between two rival techniques, which are each other's equal in all other respects, it would be hard to argue against choosing the one with the strongest history of achievement.

Individual systems may place particular demands, or impose certain restrictions, on the site, which may need to be taken into account at this stage. This is particularly relevant in respect of utility provision, where, for example, a given process may need a regular supply of large quantities of water, or have an atypically high power requirement. Though there is some evident cross-over here, this is probably best looked on as an intrinsic need, which must be met by reference to external factors. If the local conditions cannot satisfy this, then the technology has effectively ruled itself out of the running. The levels of skill and expertise required to run the plant, though a relevant consideration for the overall programme, is a less rigid criterion. Though it inevitably features in the equation, it is obviously possible to overcome any shortfalls in this area by recruitment or retraining. The costs involved in doing so may themselves prove prohibitive, but this is more properly viewed as an external issue of funding and not an unavoidable consequence of the technology itself.

The environmental benefits brought vary between different biological waste treatments and again, though their relative perceived merits will be moderated by specific local influences, it will prove necessary to make some form of direct comparison of the inherent advantages each method offers. Between individual examples of the same fundamental technology, the essential similarities may make this of little or no relevance. However, the difficulties involved in any attempt to assess the widely differing environmental gains offered by the broader spectrum of approaches, such as composting, anaerobic digestion and ethanol production, against a single, objective yard-stick are self-evident. Nevertheless, since much of the driving force behind both biowaste diversion and its public acceptance has arisen out of a desire to behave in a more earth-friendly fashion, difficult or not, some mechanism must be used to ensure that these benefits are taken into account.

Finally, there will always be the possibility that changes will occur over the duration of the project, particularly when a long lifetime is proposed. Thus, the flexibility of each technology to accommodate variations in the biowaste itself, its collection and separation arrangements, and the ease of adaptability it offers to alterations in wider waste management practice need to be considered carefully, especially if there is a long-term commitment to a more fully integrated approach.

External Factors

The scope for external influence is very large and exists across a wide range of levels, from the expressly local, through regional, to national considerations and the individual effect of each may be further modified by the interplay between them. At the top, a given country's laws or established targets have obvious bearing on any choice of technology to be applied. At the other end of the scale, a specific municipality's priorities, or some particular local circumstance, may also be enough to sway the balance in one direction or another. Consequently, external aspects may often become the deciding factor in distinguishing the right practical approach for the particular project, from the array of generally possible methods on offer. There are three main areas in which these outside influences arise, firstly as waste management issues, secondly as elements of economics and thirdly as environmental considerations, though obviously there is a measure of overlap between these classes.

Waste Management Issues

As has been discussed on previous occasions, there is a degree of variability within waste streams, in part influenced by cultural economic and similar factors at a national level, and by the localised socio-demographics of the population in individual areas. Many contributory elements can additionally shape the overall refuse profile and it is, therefore, important that any decisions on biological treatment are made on the basis of the fullest available understanding of the particular biowaste character in question. While this may not necessarily make much difference to a plant taking nothing but household or garden material, when the planned facility is expected to offer effective biological co-disposal to these components, together with additional inputs from commercial sources, the importance of such detailed knowledge increases. Not only the physical amount of the total feedstock and the individual types of biowaste within it, but also its relative compositional makeup, can have major implications for the technology needed to process it. In as much as it is essential to know that an individual technology can deal with the particular waste available, it is equally vital to ensure that the nature of the local biowaste is well enough known to make such an assessment possible.

Although, as was pointed out earlier in this chapter, the transport, handling and separation methods to be used can be viewed as largely independent of the biotechnology, the actual means of collection for a particular area can play a major part in the formulation of the biological treatment plan. The influences on final product quality exerted by mixed or separate collection regimes has already been discussed previously, along with their consequent marketing ramifications. It may be necessary to modify the collection method to permit easier biowaste diversion, or the intended biological processing approach, if the collection system cannot itself be altered, which may then have a knock-on effect in terms of how the waste is dealt with on reception at the facility. Thus, the eventual treatment technology chosen may heavily depend on the relative purity of the biowaste fraction obtained and this is dictated by the final efficacy of the sorting, performed either at the kerbside or on-site, which is itself a factor in the overall collection regime. All of this is heavily interwoven into other issues, including the employment market, local economy and available municipal funds.

It is also intimately related to the needs of any other waste initiatives or recycling schemes.

These are themselves important considerations for the biowaste plan in a variety of ways. As just discussed, existing or proposed arrangements may affect the nature of local refuse collection. They may also act to increase general awareness of waste issues or to pre-train householders into a source separation ethos. Alternatively, if a municipality has already taken the decision to travel along the MRF route, there may be valid reasons for adopting a biological treatment approach which enables them to maximise the benefits from their prior investment. The success of other local initiatives or, perhaps more importantly, how widely their successfulness is perceived, may be a factor in helping to predispose more of the population to show active support for the new scheme. There may also be some economies of scale to be gained from dovetailing the biowaste project into existing arrangements. Clearly, for any local authority looking towards truly integrated waste management as their ultimate goal, it is essential that the potential for this kind of synergy be explored as a major element in the technology selection procedure. Even for projects with less immediately ambitious objectives, the inevitable need to consider the collection, transport and financing aspects of refuse, as part of a single local waste policy, still dictates that the combined effect of any other initiatives in the area, must form a part of the overall biowaste equation.

Economic Elements

The first and most obvious economic influence on the biotechnology selection, as well as on the viability of the entire project itself, is the requirement of the local market. For example, composting may well provide the most appropriate product for an area with a particularly heavy demand for on-going land reclamation, or where there are large municipal parks and gardens to provide a ready use. In other places, perhaps keen to encourage industrial investment, the possibility of energy production and a low-cost district heating scheme may be a more attractive proposition. As has been stressed before, one of the paramount factors in the expansion of biowaste treatment will always be the establishment of bulk end user outlets and this can only ever be achieved by reference to local needs. This is an aspect which is not confined only to the discussion of large-scale, centralised facilities; it applies every bit as much to smaller, householder-based initiatives like home composting. For some areas, providing individual compost bins can be a useful approach to the biowaste issue, especially if linked to an overall strategy of minimisation or separate collections for recycling. While there is no true 'market' under such an arrangement, there may be a genuine incentive to homeowners, if they can produce and use a good quality compost. However, local conditions are essential factors for overall success; without gardens, or relying on an ineffective composter, the same scheme will flounder as the product no longer has any perceived value, as some councils have already discovered.

The influence of local culture and attitudes is also one which may need to be taken into account in the selection of the appropriate technology. Education programmes designed to raise public awareness can be useful, but it is likely that cooperation levels from different sectors of the community will vary and there will always remain areas where the take-up rate for any externally designed programme will be disproportionately low. For any initiative requiring a high degree of participation to succeed, this may prove a major obstacle.

Certain aspects of the local economy may also have a bearing on the approach taken to biowaste. The utility pricing structure, road fuel, employment costs, current waste disposal charges, land values and similar factors can all affect the relative worth of individual methods of biological treatment and ties in with the earlier caveat regarding processes developed overseas. Thus, effective techniques which are highly labour-intensive, for instance, may be eminently suitable for countries where employment costs are low, but obviously not so, where they are higher. Systems which are competitive in their home market, against expensive disposal options, are seen as far less attractive by nations where landfill is cheap and abundant. There are many similar examples and this general pattern holds true for all technologies which are heavily dependent on any such nationally variable cost for their success. While regional variations within the same country, or at the more localised state or county level, are seldom so radical, they may still impinge on the selection of appropriate technology. On the one hand, the biowaste input to the proposed facility may reflect local commercial influences, either directly, in the form of refuse from particular factories, or indirectly, as a result of particular product availability or use in the area. On the other, the commercial outlets offered to the final product from biological treatment may well depend on the industries present in the area. Even more immediately, the scope of options open to a municipality may be limited by what the local economy can support. The most obvious aspect of this is the directly financial burden imposed, but the full extent may ramify much further afield.

One of the areas that this may touch is employment. The need to assess the likely human resources required for the plant, and the possible necessity for training to be provided, has already been mentioned. For some regions, particularly those experiencing a down turn in their economy, or loss of traditional industries, the underlying employment trends may be powerful factors to consider in the overall project planning process. Within Europe, for example, such disadvantaged areas may qualify for additional funding under various of the EU programmes, especially when the outcome is thought likely to stimulate the local economy or provide new jobs. Many other countries also have their own priority areas for similar investment incentives. Eligibility for such assistance may require certain conditions to be satisfied, either in terms of the overall scope of the project, or more simply in terms of the origin of the equipment or technology to be used. Clearly, where such issues are relevant, they provide a very straightforward selection mechanism. Even for areas where they are not, the provision of new employment alone remains a very attractive proposition and it is difficult to see how this could not weigh heavily in the deliberations of any municipality or local authority. However, the quality of the work opportunities may be just as important a consideration. A number of materials recovery facilities have been proposed over the years which relied substantially on manual sorting. While some of these offered quite reasonable conditions to the workforce, others were heavily flawed, depending largely on the disenfranchised elements of society. Even in countries which view meeting their social obligations very seriously, separating mixed refuse by hand is a relatively poor job and there is, consequently, often a high turn-over rate of operatives. The issue of employment is, then, not clear cut and its influence on the overall aim of biowaste treatment depends very significantly on the wider perception of individual priorities within the locality.

The financial element is another major consideration, both in terms of the available funding for the instigation of the project and the general fiscal environment within which it must function. Obviously a municipality will have to judge the relative merits of the available biological waste treatments against the costs associated with them, together with any necessary concomitant expenditure on equipment and infrastructure required. The demands of value for money, coupled with finite budgets, can act as barriers in their own right, though various external sources of funding may be available, subject to individual circumstances, to overcome some of the limitations thus imposed. Technology selection must also be made against the backdrop of any economic instruments, such as taxation, fuel escalators, excise duties, relevant grants, preferential loans and investment incentives, which may apply to waste management in particular, or to businesses in general, in a given country or region. Such factors can often play key roles in determining the overall viability of a biowaste project and represent one of the most nationally variable elements in the whole equation. This can be deliberately manipulated to provide a financial imperative to prioritised goals, as in the case of the landfill taxes imposed in some countries. Various nations have used a steadily escalating taxation burden to alter the economic balance and force the consideration of alternative avenues of waste management. Biowaste treatment itself has been one of the major beneficiaries, particularly in those countries, like the UK, which make a rate distinction between 'active' and 'inert' waste entering landfill. Commercial taxes on businesses may also slant the relative costs of individual technologies and where councils are intending to put the provision of service out to tender, this may be an important factor.

Although at more local levels the degree of variance is not so large, nevertheless there remains scope for the regional economic environment to exert its own influence, for example, as in assistance packages to encourage foreign investment or incentive arrangements to promote the regeneration of derelict industrial sites. One thing is for certain, biological waste treatment must be looked on as an essentially commercial activity and, consequently, it occupies a world governed by the Best Practicable Environmental Option (BPEO) and the Best Available Techniques Not Entailing Excessive Cost (BATNEEC). Thus, economic aspects will always be large influences on the uptake of all biowaste initiatives and, most particularly, in the selection of methods to be used in any given situation. It is impossible to divorce this context from the decision making process.

Environmental Elements

Mention of BPEO leads naturally to the final category of external factors, which relate to the environment. It is obvious that the technology selected should sit comfortably alongside the wider aim of responsible environmental management. It is, therefore, essential that the chosen approach be suited to the local situation, and the absolute minimum requirement should be that it will not contribute to polluting, or otherwise despoiling, the area. Clearly, those methods which bring actual benefits or improvements will generally be seen as preferable. Though there is a broad consensus about the major issues of environmental protection, often there is disagreement as to how they may best be addressed and this may lead to a number of conflicting viewpoints emerging during the discussions of the appropriate technology route. For example, as discussed previously, it is possible to make two quite different cases for paper use, or to argue that the best practicable environmental option for food waste, at least in some areas, consists of macerating it in waste-disposal units, and consigning it to the water company for treatment. There are many more examples where opinions may vary on the best means by which to proceed, and often it comes down to a final decision based on the environmental equivalent of trying to compare chalk with cheese. In the wider context, the individual priorities of a nation, or region, in respect of the reduction of fossil fuel usage, atmospheric carbon dioxide contribution, reliance on landfill or incineration and so on, will inevitably tend to colour the way in which the ultimate selection is made.

The question of local benefits is also one which may carry much weight for some areas, particularly where a better waste management service can be offered to the population at the same time as some additional amenity value gained. This can be a particularly persuasive argument when the scope of the proposed biowaste project is such that it is intended to bring in material from neighbouring districts to be processed. While the additional revenue and kudos such an arrangement provides is of obvious benefit to the authority responsible, care must be taken that this move is not simply perceived as the area becoming a dumping-ground for everybody else's waste. Understandably, councils can be very sensitive about this, particularly if there is a historical record of polluting industrial development associated with the region and the prospect of some visible gain, be it in the form of a visitor centre, district heating or good publicity, can be a very powerful incentive.

On a more local basis still, aspects of the availability and usage of land may be some of the most relevant environmental factors to consider. In as much as the technologies themselves may make certain demands on their proposed location, the site itself can impose restrictions on the methods which may be used. As mentioned at the outset, these influences typically apply irrespective of whether a particular piece of land has already been firmly identified, though, obviously, if a municipality has no choice in the siting of its facility, then their effect is even more apparent. Prior activities at the location itself, its topography, geology and hydrogeology, the existing neighbours and its proximity to other sensitive sites, judged on either ecological or human grounds, may all place constraints, of one form or another, on its proposed new use. Such considerations may limit the available technology options and if no alternative venues are available, they may, as part of the relevant procedures for planning or site approval, provide the final defining boundaries for the whole project.

Monitoring and Facility Performance Criteria

In order to be able to gauge the success of the biowaste initiative in operation, some programme of on-going monitoring is typically required, and it is often helpful to draw this up against a background of pre-defined performance indicators, which make evaluation and analysis more clear-cut. Effective monitoring can be seen as fulfilling three basic goals:

1. Ensuring environmental compliance
2. Forming a process control and management system
3. Providing a predictive or modelling tool

Each of these three may be further subdivided, to form the following key areas:

i   On Site Considerations
    This encompasses vehicle movements, noise, wind-blown litter and the immediate effects of the operation.

ii The Products
    Covering the likely impact of derived products and their final uses, this includes 'waste' products, like exhausted process liquor, as well as beneficial materials. Effective monitoring here enables a measure of quality awareness, which is essential for market placement and to assist in any eventual requirement for cleansing.

iii Normal Operational Considerations
    Charting the routine running of the facility, to give valuable information as to the efficacy of the particular process used and the general progress being made.

iv Extraordinary Considerations
    Data collection for the prediction and early warning of accidents or emergencies, which could also be used to suggest effective remediation measures in these circumstances.

However, for many actual operations, these topics are often more usefully considered in respect of how they relate to the specific process used, the final product and the environment, since these issues form the main practical focus of the monitoring regime. As will be readily appreciated, these remain essentially artificial distinctions and there is, consequently, some measure of cross-over between them. From a more pragmatic standpoint, the wider outcomes from monitoring include the efficient management of the process and running of the facility, the evaluation of any pollution or environmental impact arising, examining public perceptions of the project and enabling effective forward planning to deal with future needs.

Process monitoring will be largely determined by the choice of technology itself, with each system having its own key indicators to be considered, this holding true also for the output, and any by-products arising. By using a properly designed sampling and analysis plan, the progress of the actual biowaste treatment can be followed and the quality of the final product can be assessed. Another aspect needing to be addressed alongside this is the equipment, since most operations have a degree of reliance on machinery of some sort or another. Although the types of devices again vary with the technology and its level of sophistication, generally they can represent a major influence on system efficiency and, consequently, their performance needs to be taken into account. The question of the facility's general management is to be judged against other criteria, typically using economic measures such as cost–benefit analysis, to ensure that any other performance targets which may have been set, are met. Since it is generally the case that the overall performance is regulated by the influx of suitable material and the demand for the derived product, plant 'efficiency' is something of a hybrid concept; a compromise between the absolute scientific quantification of the process and elements of the relevant financial model, as thus applied.

The need for pollution or emission monitoring is self-evident. The potential contamination issues will be dependent on the particular method of treatment employed and the potential pathways and targets likely to be affected will also vary as a result of site-specific factors. For some facilities, it has been found more convenient to deal with acute pollution issues as part of the routine process monitoring scheme, while rolling the longer-term impact studies into a general environmental audit process, though this may not be an appropriate strategy for all operations.

The issue of public perception, as has been mentioned on various occasions throughout this book, is one of extreme relevance to the success or failure of any biowaste treatment initiative. At every step of the way, from the initial planning, through actual use, to providing, in many cases, an outlet for the product, the active support of the local population is essential. Moreover, it is clear that any intended expansion of the project can only be realised if there is an established pattern of use by those for whom it was first set up. In order to achieve this, it is important that the facility, and the way it is run, meets the legitimate needs of the people and is, furthermore, perceived as beneficial by them, from the outset. Any reluctance in the initial stages of the scheme's lifetime may have serious repercussions on its long-term viability and so it has frequently proved necessary to view these projects in terms of a series of distinct phases, with a rolling programme of education and gradual change, to gently bring about the desired alteration in outlook. Only with some form of monitoring of this aspect of the site's usage, offering the users an opportunity to provide feed-back based on their experiences, can actions be taken to try to ensure their continued support. This is principally of use to the likes of intermediate scale operations, where the public supply their segregated biowaste directly, and there is some evidence to suggest that, particularly where such schemes are co-located with other recycling facilities, this is where the greatest confusion is likely to arise over the correct placing of waste types. However, the matter is relevant to larger initiatives also and the results obtained in all cases should receive due consideration.

Both waste and technology alter over time, though generally, the former, slower than the latter. For any system intended to enjoy a relatively long operational lifespan, there is an inbuilt necessity for a certain degree of flexibility to accommodate these changes, either in terms of the advent of better equipment and techniques, or in the composition of the waste itself, as has been discussed. One of the best approaches for coping with the needs of the future is through effective prediction and adequate forward planning, both of which rely on good information gathering at the earliest opportunity. One of the problems highlighted in earlier sections of this book is the frequent lack of comprehensive waste analysis data available for planning purposes, particularly in respect of precisely this type of variation over time. The circumstances are ideal for ensuring a solution to this problem when a new facility is established, as the necessary procedures to obtain these requisite details can simply be set up to run alongside the rest of the newly implemented monitoring regime, permitting seamless integration with the broader aspects of process evaluation. Hence, with an accurate supply of information, trends can be foreseen with some degree of certainty and any appropriate alterations to the scope or manner of the biowaste provision can be made with greater confidence. However, the value of this particular aspect of monitoring is not solely restricted to considerations of restructuring the system itself. It can also provide the most direct means to gauge the way the facility is used, the efficacy of measures designed to alter public perception, and, by implication, its usefulness to the local community.

Evaluation and Analysis for Client and Contractor

A good number of the local authorities charged with responsibility for waste involve private sector companies in the provision of biowaste facilities for their area. This can take the form of anything from initial consultancy, through design and built packages, to a completely contracted out service. Many of the rest view the relationship between the executive and their own operational arm as one of client and contractor also. Hence, it is often convenient for both sides to have access to an effective means of evaluation and analysis of the project's overall performance, although their particular criteria for success may differ. For example, in biological processes involving the generation of a saleable product, both parties might be interested in the quality, amount and consistency of material produced. However, the same characteristics of the incoming biowaste would probably only be relevant to the contractor, especially if there is some clause in the contract to govern the state of the delivered refuse, which is often a feature where the biowaste operator has no control over the refuse separation.

Such evaluation relies heavily on the gathering of appropriate data and its eventual statistical analysis, to yield meaningful information about the project, the way it is run and how it meets the targets set for it. Performance indicators may be imposed and compliance judged in this way, though it may sometimes be necessary to view the overall effect of the facility in wider terms than simply on the basis of a mass-flow through its gates. Thus, for example, the efficacy of a biowaste treatment site as a diversionary strategy would have to be determined by reference to the total landfill figures, as well as in respect of the tonnages arriving at the plant itself. The overall monitoring process can be particularly valuable in identifying trends and characteristics in the performance of the treatment process. However, adequate record keeping is an essential element, since the information obtained may be of considerable significance for many aspects of future planning. An organised and systematic method of documentation, whether or not actually required by law or the client, must be regarded as indispensable, if for no other reason than for their direct relevance to the eventual long-term evaluation of the whole project.

The inter-relationship between all of the interested parties is critical to the success of any facility processing biowaste and this goes beyond the simple client and contractor framework. In the final analysis, the constructive cooperation between government, regulatory bodies, municipalities, private sector companies, the workforce and the public stands as a major factor in the expansion of biological waste treatment. It is imperative that all of these groups maximise the coordination of their respective approaches, requirements and activities to avert needless conflict or unnecessary duplication of work. This may be impossible where statutorily imposed duties force the overlapping of responsibilities between different organisations, but elsewhere, the removal of any avoidable procedural repetition would be a significant move. Moreover, such close cooperation, particularly on a regional basis, is a vital element in ensuring the continued local relevance of the programme, which, as has already been discussed, is itself important in guaranteeing long-term community support for the project.

Biowaste Treatment and Recycling Obligations

Around the globe, recycling targets are becoming increasingly common and the pressure to achieve them mounts, driven by the requirements of ever more stringent legislation, supported by financial penalties or incentives. The issue of resource management, virtually unheard of, even up until only a few years ago, has begun to be recognised by a steadily growing number of people. Recycling itself, if not yet second nature to all, is at least now familiar to most. Against this background of enhanced awareness, however, and despite a stated EU policy to stabilise waste generation at 1985 levels[1], waste volumes in most European countries have continued to grow. This is a pattern which has been repeated elsewhere in the world. The full implications of this are not easily determined, as a result of the lack of comprehensive historical data, mentioned elsewhere in this book, and an incomplete understanding of the interactions between all the factors which influence refuse production in general. Decisions regarding the provision of necessary facilities must, therefore, be made on the basis of a prediction of the relative effects of those factors which lead to the increase in waste production, against the counter influence of social and legislative pressures to reduce it. Many aspects of life alter the amount and kind of refuse arising and the technological advances and continued changes to working patterns probable in the 21st century are likely to continue the trend, making flexibility and robustness equal requirements of any long-term waste strategy. Demographic shifts and the upsurge of single-occupant households have also been implicated as possible contributors to the additional total, while the environmental impact of moving waste long distances for treatment or disposal has also begun to feature more prominently in discussions of policy.

It has come to be widely recognised that disposing of some products when they have served their immediate purpose, may actually represent a wasted opportunity in itself. To many people, the question of how some form of value may be derived from what is thrown away is best answered by looking towards recycling, to garner all the useful materials from the waste stream and give them a second chance in the chain of utility. Certainly, metals, plastics, glass and paper may all be reused, in one way or another, to make a useful product and thus the whole idea has become a very popular one, synonymous with all that is good and right and 'green'. But is this, alone, enough?

Municipal solid waste accounts for a relatively small amount of the annual total refuse produced and yet it is, perhaps, one of the most problematic to deal with, by virtue of, firstly, its compositional heterogeneity and, secondly, its high biodegradable content. However, while the biowaste element may present a pollution risk, especially under landfill conditions, it also offers an additional, and significant, potential increase to the overall recycling rate. Such bio-cycling adapts the mechanisms of the natural carbon and nitrogen cycles very easily to achieve the controlled reclamation of nutrient and fibre for beneficial use. Thus, the biological material discarded by man, food waste, grass cuttings, paper and the like, may be reprocessed and returned to the soil to support another round of production and consumption. While, for biowaste, it may be difficult to see minimisation initiatives having a major effect, the other hierarchical goals of sustainable waste management remain relevant and it is possible to make the best use of the discarded putrescible material, and reduce the likelihood of pollution, in this way.

Biological waste treatment, like traditional recycling, is typically more labour-intensive, on a tonne for tonne basis, than other forms of waste management and so any upsurge in landfill diversion is likely to have implications for the wider community. Particularly in the more densely populated areas, extra employment opportunities may be created as the local collection, separation and reprocessing industries expand to meet the newly created demand. However, the long-term prospects of these remaining permanent depend on a number of economic conditions and it follows that any external subsidy introduced to drive higher rates of recycling will play its part in deciding this. Attitudes to such artificial fiscal intervention vary between countries and many take the view that regional self-sufficiency is of paramount importance in respect of resources management. This approach is not without its advantages, especially since it is becoming widely recognised that the requirements of waste policy, generally, do not always fit particularly easily within the strict confines of individual local authority borders. All aspects of waste management have, thus, been described as 'transboundary issues'[2] and this serves as a reminder that there are limits on what any municipality can realistically be expected to achieve on its own.

Nevertheless, recycling targets ultimately tend to rest on individual authorities to achieve and some of the levels at which they are set will make stern demands on the available resources, in order to meet them. If waste management is truly to become 'integrated' or 'sustainable' over the coming years, then the need for the kind of close cooperation between various relevant bodies discussed previously, will grow and the case for a raft of seamlessly interlinked, mutually supportive technologies will effectively make itself. It has been said that the distinction between a biologist and a physicist is that, while the former looks for differences, the latter is concerned with similarities[3]. Perhaps, then, in this discussion of the biological, we must play the physicist, for once. It will always be important, for the reasons set out earlier in this chapter, to keep sufficient flexibility to allow individual biowaste solutions to be implemented in particular areas. However, in overview, the differences between the various biotechnologies, and even between processing biowaste and other kinds of waste recovered materials, are far less important than their similarities. In the final analysis, perhaps the really meaningful distinction is between disposal and recycling.

Leaving aside any diversion requirement, with around 60% of European household waste streams being 'biodegradable', as defined by the EU, it is clear that an efficient biological recycling approach could offer valuable help towards reaching the increasingly high recovery targets being set. Indeed, given this preponderance of biowaste, it is hard to imagine how any authority could envisage meeting its recycling obligations, without some recourse to biological waste treatment.

References

1. Commission of the European Community, *Towards Sustainability: A European Community Programme of Policy and Action in Relation to the Environment and Sustainable Development* Council of Ministers (92)23 Vol.2, March 1992.
2. Rogalski, W. *Status and Trends for Biological Treatment of Organic Waste in Europe*, International Solid Waste Association, Denmark, 1995, p. 4.
3. David Moreland, the author's A Level physics teacher.

Contacts

Waste Management Organisations

International Solid Waste Association
(ISWA)
General Secretariat
Overgaden Oven Vandet 48 E
DK-1415 Copenhagen K
Denmark
Tel: +45 32 96 15 88
Fax: +45 32 96 15 84
Email: iswa@inet.uni2.dk
www.iswa.org

ISWA Austria
11. Haidequerstrasse 6
1110 Wien
Austria
Tel: +43 1 760 99 112
Fax: +43 1 760 99 316
Email: iswa@iswa.at.

DAKOFA
Overgaden Oven Vandet 48 E, St.
DK-1415 Copenhagen K
Denmark
Tel: +45 3296 9022
Fax: +45 3296 9019
Email: dakofa@dakofa.dk
www.dakofa.dk

Association Gènérale des Hygiénistes et
Techniciens Municipaux
(AGHTM)
83 Avenue Foch - B.P. 39.16
75761 Paris - Cedex 16 - France
Tel : +33 (0)1 53 70 13 51 ou 53
Fax: +33 (0)1 53 70 13 40
Email: aghtm@aghtm.org
www.aghtm.org

NVRD
Postbus 1218
6801 BE ARNHEM
The Netherlands
Tel: 026-377 13 33
Fax: 026-445 01 55
Email: post@nvrd.nl
www.nvrd.nl

Solid Waste Association of North America
SWANA
P.O. Box 7219
Silver Spring
MD 20907-7219
USA
Tel:1-800-GO-SWANA (1-800-467-9262),
Fax: +1 (301) 589-7068,
Email: info@swana.org
www.swana.org

RVF - Svenska
Renhållningsverksföreningen
Östergatan 30
211 22 Malmo
Sweden
Tel: 040-35 66 00.
Fax: 040-97 10 94
Email: office@rvf.se
www.rvf.se

Institute of Wastes Management
IWM
9 Saxon Court
St Peter's Gardens
Northampton NN1 1SX
UK
Tel: +44 (0)1604 620426
Fax: +44 (0)1604 621339
Email: technical@iwm.co.uk

membership@iwm.co.uk
education@iwm.co.uk
www.iwm.co.uk

National Solid Wastes Management
Association (NSWMA)
4301 Connecticut Avenue, NW
Washington DC 20008
USA
Tel: +1 202 - 244 47 00
Fax: +1 202 - 966 48 41
Email: eii@envasns.org
www.envasns.org/nswma/

Biomass Fuels

American Bioenergy Association
1001 G Street NW, #900 East
Washington DC 20001
USA
Tel: +1 202 - 639 03 84
Email: info@biomass.org
www.biomass.org

British BioGen
7th Floor,
63-66 Hatton Garden
London
EC1N 8LE
UK
Tel:+44 (0)1207 831 7222
Fax: +44 (0)207 831 7223
Email: web@britishbiogen.co.uk
www.britishbiogen.co.uk

Canadian Association for
Renewable Energies
435 Brennan
Ottawa   K1Z 6J9
Canada
Tel: +1 613 - 728 08 22
Fax:+1 613 - 728 25 05
Email: eggertson@renewables.ca
www.renewables.ca

Gasification Technologies Council
1110 North Glebe Road, #610
Arlington   VA 22201
USA
Tel: +1 703 - 276 01 10

Fax: +1 703 - 276 76 62
Email: info@gasification.org
www.gasification.org

Groupe Energie Biomasse
Université Catholique de Louvain
Place du Levant 2
1348   Louvain-la-Neuve
Belgium
Tel: +32 10 - 47 83 98
Fax: +32 10 - 45 26 92
Email: geb@term.ucl.ac.be
www.meca.ucl.ac.be/term/geb/

National Biodiesel Board
PO Box 104898
1907 Williams Street
Jefferson City   MO 65110
USA
Tel: +1 573 - 635 38 93
Fax: +1 573 - 635 79 13
Email: info@biodiesel.org
www.biodiesel.org

Talbott's Ltd
Drummond Road
Astonfields Industrial Estate
Stafford
ST16 3HJ
UK
Tel: +44 (0)1785 213366
Fax: +44 (0)1785
Email:
www.talbotts.co.uk

Energy

Centre for Energy Policy
Petrovka 14
103031   Moscow
Russian Federation
Tel: +7 95 - 200 37 34
Fax:+7 95 - 200 44 79
Email: webmaster@energy.ru
www.energy.ru

Center for Renewable Energy &
Sustainable Technology. (CREST)
1200 18th Street NW, #900
Washington DC 20036
USA

Tel: +1 202 - 530 22 30
Fax: +1 202 - 887 04 97
E-Mail: info@crest.org
www.solstice.crest.org

Compost / Soil Products
Composting Council
4424 Montgomery Ave, #102
Bethesda   MD 20814
USA
Tel: +1 301 - 913 28 85
Fax: +1 301 - 913 91 46
Email: ComCouncil@aol.com
www.compostingcouncil.org

Henry Doubleday Research Association
Ryton Organic Gardens
Ryton-on-Dunsmore
Coventry
CV8 3LG
UK
Tel: +44 (0) 1203 303517
Fax: +44 (0) 1203 639229
Email: enquiry@hdra.org.uk
www.hdra.org.uk

Swedish University of Agricultural Sciences
Department of Resource & Environmental Economics
PB 7013
750 07   Uppsala
Sweden
Tel: +46 18 - 67 17 51
Fax: +46 18 - 67 35 02
Email: Andrew.Dragun@ekon.slu.se

General Recycling
Bureau of International Recycling
24 Avenue Frankin Roosevelt
1000   Brussels
Belgium
Tel: +32 2 - 627 57 70
Fax: +32 2 - 627 57 73
Email: info@bir.org
www.bir.org

Sustainable Development
Ekocentrum Gothenburg
Sankt Jorgens vag 20

422 49   Hisings Backa
Sweden
Tel: +46 31 - 55 68 82
Fax: +46 31 - 55 68 81
Email: ekocentrum@ekocentrum.nu
www.ekocentrum.nu

International Institute for Environment and Development (IIED)
3 Endsleigh Street
London   WC1H 0DD
UK
Tel: +44 171 - 388 21 17
Fax: +44 171 - 388 28 26
Email: mailbox@iied.org
www.iied.org

Sustainable Development for Local Authorities
Van Eeghenstraat 77
1071   Amsterdam
The Netherlands
Tel: +31 20 - 578 76 00
Fax: +31 20 - 662 23 36
Email: eva.klok@imsa.nl

Sustainable Development Networking Programme
CGO Complex
Lodi Road
Paryavaran Bhawan
110 003   New Delhi
India
Tel: +91 11 - 436 21 40
Email: sdnp@envfor.delhi.nic.in
sdnp.delhi.nic.in

Environmental Management
Universal Development Environment Centre
No.80, Weera Mawatha
Bokundara
10300   Piliyandala
Sri Lanka
Tel: +94 1 - 61 93 65
Fax: +94 1 - 84 32 76
Email: analytica@eureka.lk

Environment Management And
Information Liaison
Postbus 3010
2301 DA   Leiden
The Netherlands
Tel: +31 71 - 523 06 52
Fax: +31 71 - 523 06 83
Email: e.m.a.i.l@club.tip.nl
www.email.club.tip.nl

Environmental Resource Center
(ERC)
1010 Teston Road
Maple, ON   L6A 1E9
Canada
Tel: +1 905 - 770 88 28
Email: sengel@idirect.com

Wales Environment Centre
QED Centre
Treforest Estate
Pontypridd   CF37 5YR
UK
Tel: +44 1443 - 84 40 01
Fax: +44 1443 - 84 40 02
Email: wec@arenanetwork.org
www.arenanetwork.org

Wastec
4301 Connecticut Avenue NW, #300
Washington DC 20008
USA
Toll-Free
Tel: +1 202 - 364 37 07
Fax: +1 202 - 966 48 24
www.wastec.org

Air & Waste Management Association
3/F, One Gateway Center
Pittsburgh   PA 15222
USA
Tel: +1 412 - 232 34 44
Fax: +1 412 - 232 34 50
Email: info@awma.org
www.awma.org

Government Agencies / Depts

Environmental Protection Agency
Office of Intl. Activities
401 M Street SW, #A-106
Washington   DC 20460
USA
Tel: +1 202 - 260 30 08
Fax: +1 202 - 260 78 75
Email: webmaster@epamail.epa.gov
www.epa.gov

European Environment Agency (EEA)
Kongens Nytorv 6
1050   Copenhagen K
Denmark
Tel: +45 33 - 36 71 00
Fax: +45 33 - 36 71 99
Email: eea@eea.eu.int
www.eea.eu.int

Bundesamt für Umwelt, Naturschutz und
Reaktorsicherheit
Referat Öffentlichkeitsarbeit
Postfach 120629
53048   Bonn, Germany
Tel: +49 228 - 30 50
Fax: +49 228 - 305 32 25
Email: oea-1000@wp-gate.bmu.de
www.bmu.de

Agence de l'Environnement et de la
Maîtrise de l'Énergie (ADEME)
27 Rue Louis Vicat
75737   Paris 15, France
Tel: +33 1 - 4765 20 00
Fax: +33 1 - 4645 52 36
Email: webmaster@ademe.fr
www.ademe.fr

Environmental Management Bureau
6/F, Philippine Heart Centre Building
East Avenue, Diliman
Quezon City
Metro Manila   3008
Philippines
Tel: +63 2 - 97 56 98
Facsimile +63 2 - 97 32 54

Ministerio do Meio Ambiente (MMA)
Secretaria Executiva
Bloco B
Esplanada dos Ministérios

Contacts

70.068-906 Brasília, Brazil
Tel: +55 61 - 322 82 25
Fax: +55 61 - 322 82 15
Email: informma@mma.gov.br
www.mma.gov.br

Ministry of the Environment
PO Box 10-362
84 Boulcott Street
Wellington
New Zealand
Tel: +64 4 - 473 40 90
Fax: +64 4 - 471 01 95

Ministry for the Environment
Vonarstraeti 4
150 Reykjavik
Iceland
Tel: +354 560 96 00
Fax: +354 562 45 66
Email: tryggvi.felixson@umh.stjr.is

AEA Technology
Waste Management Information Bureau
Abingdon OX14 3DB
UK
Tel: +44 1235 - 46 33 91
Fax: +44 1235 - 43 29 16
www.aeat-env.com

Environment Agency (UK)
General Enquiry Line
0845 9333111

National Environmental Information
Service (NEIS)
3154-B College Drive
Baton Rouge LA 70808
USA
Tel: +1 505 - 377 12 24
Fax: +1 505 - 377 12 25
Email:mcouhig@ix.netcom.com
www.neis.com

World Health Organization (WHO)
20, Avenue Appia
1211 Geneva 27
Switzerland
Tel: +41 22 - 791 21 11
Fax: +41 22 - 791 07 46
Email: postmaster@who.ch
www.who.org

Index

Acclimatisation 59, 111
Acetic acid 91
Acetogenesis 6, 91, 94
Acidogenesis 90–1, 94
Actinomycetes 61–2
Aerated pile 70
Anaerobes (obligate) 91
Annelidic conversion 120–123, 158

Bacteria
    - *In AD* 92–94, 96–7
        - *Archaebacteria* 93
    - *In composting* 60–62
Best Available Techniques Not Entailing Excessive Cost (BATNEEC) 179
Best Practicable Environmental Option (BPEO) 19, 36, 155, 165–6, 179
Bioaerosols 49, 84
Biofuel 136
Biogas 95, 109–111, 118, 144
BOD 4, 6–7, 113, 128–9
Butyric acid 97

Carbon dioxide 7, 19, 90–1, 94–96 101, 118, 143, 147, 167
    - *Properties of* 95
Carbon: Nitrogen ratio 4, 60, 74, 76–7
Cation exchange capacity 114
Cellulose 3, 61, 90, 96, 124
Certificate of Technical Competence (COTC) 23
CES Oxynol 124
Circum-phyllospheric micro-organisms 80
Closed reactor 70
COD 4, 6–7, 113, 128–9
Compost accelerants 75–6
Compost teas (infusions) 80–1
COSHH 21
Combined Heat and Power (CHP) 15, 137, 144–5
Compost biology 61–63
Cryophilic 90

Digester types 104–5

Endotoxins 54, 84-5
Energy production 116, 124–5, 136, 141, 143-5, 148, 167
Enzymes 52, 59, 90, 124, 126–7
Environmental Protection Act 10, 18, 21–2, 151
Ethanol production 124–126,136
Eutrophic fermentation 126–133
Exothermic decomposition 58

Food waste 36, 67, 73
Forced aeration 69

Gasification 143
Glucan 84
Grass clippings 4, 65

Health and Safety at Work Act 21
Health issues 46, 84–85, 142
    - *Allergic responses* 49, 85
    - *Aspergillosis* 49
    - *Fungal spores* 49
    - *Organic Dust Toxic Syndrome* 84
Hemicelluloses 3
Home composting 63–66
Homogeneity 39, 59, 68, 99
Hydrogen, role in AD 96–97
Hydrolysis 59, 90, 94
    - *Acid* 124
Hydropulpers 99

ICRCL 43
Indicator organisms 48
In-vessel composting 70
Ion interaction (in AD) 102

Lactic acid 91, 97
Landfill Directive 5, 24–5, 27, 29, 36, 41, 66, 114, 137, 148, 150–1, 159
Leachate 5–8
Lignin 3, 61, 99

Mass balance 107
Maturation 60, 76

Index

Mesophilic
- *In AD* 90, 98, 103–4, 106
- *In composting* 60, 61, 72, 90

Metals
- *Cadmium* 43–5, 78, 102, 108–9
- *Chromium* 43–5, 78, 102, 108–9
- *Copper* 43–5, 78, 102, 108–9
- *"Heavy"* 28, 29, 42–45, 51, 114, 122
- *Lead* 43–5, 78, 102, 108–9
- *Mercury* 43–5, 78, 108–9
- *Nickel* 43–5, 78, 108–9
- *Toxicity (in AD)* 102
- *Zinc* 43–5, 78, 102, 108–9

Methane 7, 90–1, 94–5, 118, 136, 148
- *Properties of* 95

Methanogenesis 6–7, 90–2, 94, 97, 100, 158
- *Rate of* 97

Microbial inoculant 62, 80, 111
Microbiological profile 61
Moisture content in composting 73
MRF 72, 84–5 138, 142, 161–162
- *"Dirty"* 38, 68, 99, 119, 132, 138, 142, 153, 161–162
MSW 3–5, 8, 10–14, 16, 22, 25–7, 34, 36–9, 41, 46, 50–52, 54, 63, 67, 70, 108, 124, 139, 141, 153, 157–61, 184
Mulch 41, 77–8, 142

National Household Waste Analysis Programme 14
Nitrification 60, 62, 76
- *Nitrate* 60, 76, 79
- *Nitrite* 60
- *Nitrobacter* 60
- *Nitrosomonas* 60

Nutrients 2, 3, 51, 58, 109, 114, 132, 146–7
- *NPK* 31, 61, 114

Optional derogation 24, 151
Oxygenation 58, 66, 74
- *Oxygen transfer* 59, 127

Paper 139–142, 153, 158
- *As biomass* 142, 166
Part 503 Rule 27–9, 47
Pathogens 5, 33, 34, 42, 46, 48, 51, 59, 78, 80–1, 122–123
- *Aspergillus fumigatus* 46
- *Entamoeba histolytica* 46
- *Escherichia coli* 46
- *Faecal coliforms* 48
- *Faecal streptococci* 48
- *Infective parasitic ova* 48
- *Legionella* 46
- *Salmonella* 46, 48
- *Micromonospora* 46
- *Thermal inactivation of* 46–7

Pesticides 50, 51
- *Fungicides* 50
- *Herbicides* 50, 65
   - *2,4-D* 51, 65
   - *Dicamba* 65
   - *MCPP* 65
   - *Pendimethalin* 51
- *Insecticides* 50
- *Methyl bromide* 80, 81
- *Organochlorides* 51
- *Persistence of* 65

pH 4, 6, 113, 129
- *In AD* 90–1, 100–101

Phosphate 76
Photosynthesis 136, 149
Phytotoxicity, 62
Pollution, 5–8, 112, 141, 146, 182, 184
Process liquor
- *In AD* 111–115,118
- *In EF* 132
Proprionic acid 91, 97
Proximity principle 19, 157, 165, 166
Pseudomonads 62
Public perception 17, 34, 182
Pyrolysis 136, 142–4

Rotary drum 70

Sanitisation 30, 33–4, 47–9, 73
Screening 71, 72
Short Rotation Coppicing (SRC) 140, 144–149, 158
Specific Oxygen Uptake Rate (SOUR) 33, 76
Stabilisation 4, 32–3, 52, 63, 76, 100, 122–3
Static pile 69
Sterilisation 59
Sulphur oxidising bacteria 61
Sulphur reducing bacteria 110
Surface area to volume ratio 39, 74
Sustainability 139, 164–5, 166

Take-up rates 52–3, 161
Thermal recycling 133, 138, 149

Thermophilic
- *In AD* 90, 98, 103–4, 106
- *In composting* 59–61, 72, 123
Total Organic Content (TOC) 7, 24, 26
Treatment trains 122–123, 133, 158
Trommel 72
Tunnel composting 69–70

Vasicular Arbuscular Mycorrhizae (VAM) 81
Volatile Fatty Acids (VFA) 7, 90–1, 96–97, 100–101, 129

WAMITAB 23
Waste classification 10–14, 160

Waste derived composts/products 30, 41, 44, 51, 78–79
- *applications* 78, 82–4 109–110
- *maturity* 62
- *plant disease suppression* 80–82, 109, 147
- *quality* 40, 50, 62, 76, 165
- *water holding* 83, 145–146
Waste derived fuel 141
Waste minimisation 162–4
Windrow 68, 74
Worm composting 120–3, 158

Yard waste 36